Wavelet Methods in Mathematical Analysis and Engineering

Series in Contemporary Applied Mathematics   CAM

Series in Contemporary Applied Mathematics CAM 14

Wavelet Methods in Mathematical Analysis and Engineering

editors

Alain Damlamian
Université Paris XII, France

Stéphane Jaffard
Université Paris XII, France

Higher Education Press

World Scientific

NEW JERSEY · LONDON · SINGAPORE · BEIJING · SHANGHAI · HONG KONG · TAIPEI · CHENNAI

Alain Damlamian
Université Paris XII
Avenue du Général De Gaulle
94010 Créteil Cedex, France
Email: damla@univ-paris12.fr

Stéphane Jaffard
Université Paris XII
Avenue du Général De Gaulle
94010 Créteil Cedex, France
Email: jaffard@univ-paris12.fr

Editorial Assistant: Chunlian Zhou

Copyright © 2010 by
Higher Education Press
4 Dewai Dajie, Beijing 100120, P. R. China and
World Scientific Publishing Co Pte Ltd
5 Toh Tuch Link, Singapore 596224

ISBN 13: 978-981-4322-86-7
ISBN 10: 981-4322-86-5

Preface

The texts of this book grew from an ISFMA (Sino-French Institute of Applied Mathematics) symposium on "Wavelet Methods in Mathematical Analysis and Engineering" that took place in August 2007 on the Zhuhai campus of Sun Yat-Sen University. This symposium was composed of a one week summer school mainly directed towards Chinese PhD students and postdocs, followed by a one week conference attended by researchers from all over the world. This event was co-organized by Sun Yat-Sen University in Guangzhou and the ISFMA in Shangai. The purpose of the courses was to give the students the required level in wavelet analysis and in the main applications that would be treated during the conference, so that they would be able to follow it with profit. The courses were given by Albert Cohen, Daoqinq Dai, Stéphane Jaffard, Lixin Shen, Zuowei Shen and Lihua Yang; they covered basic materials concerning construction of properties of wavelet bases, and alternative decomposition methods; they gave an overview of the main applications in the numerical analysis of PDEs, and signal and image processing. The workshop exposed new techniques such as Empirical Mode Decomposition (EMD) and new trends in the recovery of missing data, such as compressed sensing, and a sample of a few recent key applications of wavelets in several scientific areas. This event was part of a long term collaboration between Chinese, Singaporean and French mathematicians in the area of wavelet analysis, and a second event took place one year later, with the "Chinese-French-Singaporean Joint Workshop on Wavelet Theory and Applications" in Singapore (June 2008).

These texts essentially correspond to the courses that were given during the summer school. Put together, they give a comprehensive overview of both the fundamentals of wavelet analysis and related tools, and of the most active directions of applications that developed recently. They offer a state of the art in several active areas of research where wavelet ideas, or more generally multiresolution ideas have proved particularly effective.

The paper by Jianfeng Cai, Raymond Chan, Lixin Shen and Zuowei Shen deals with the practical problems of high resolution reconstruction of noisy and blurred images, and the super-resolution challenge, which consists in using *a priori* information on the structure of the image in order to reconstruct it at a higher resolution than is available; this ill-posed problem is tackled using multiresolution ideas, which were at the very origin of wavelet techniques, in the 1980s. The resolution of these questions is obtained through the use of tight-framelets; the paper first gives a tutorial on frames and tight frames, which are now an important tool in signal and image processing, and then focuses on tight-framelets, a tool which is exposed in details, and whose efficiency is demonstrated in this context.

The paper by Albert Cohen addresses a fundamental problem in approximation theory: how to approximate a piecewise smooth function, in a numerically efficient way by few simple "building blocks". One key idea developed in the paper is that one should use nonlinear approximation techniques: one picks the approximation from a set of functions depending on N parameters (but which does not form a vector space), and the paper starts by a tutorial on N-term approximation. For the specific image processing problem which is proposed, the author developed a particularly effective method where the building blocks are piecewise polynomial functions on triangles on which no shape restriction is imposed. This extra flexibility has the advantage of offering efficient reconstruction algorithms for functions with edges along smooth lines. The algorithms used develop refinements techniques based on multiresolution ideas. The paper demonstrates the accuracy and the numerical simplicity of the method.

The paper by Stéphane Jaffard, Patrice Abry, Stéphane G. Roux, Béatrice Vedel and Herwig Wendt gives a brief tutorial on constructions of wavelet bases, and the characterization of function spaces in terms of wavelet coefficients. It shows the relevance of these techniques in a problem posed by the seminal papers of Kolmogorov in turbulence in the 1940s, where he advocated the study of some quantities which were expected to be scaling invariant, and fundamental for the comprehension of small scale turbulence. The study of these quantities rewritten through a wavelet expansion yields unexpected new tools for signal and image classification and for the selection of turbulence models. Applications to multifractal analysis are given, i.e. for the estimation on the size of the sets of points where a function has a given pointwise Hölder regularity.

The paper by Chaochun Liu and Daoqing Dai addresses the question of face recognition. The difficulty of the problem arises form the fact that a face can change widely due to variations in pose, expression and illumination. The challenge is to find attributes that remain stable under such variations. This is precisely supplied by wavelet techniques, which yield an efficient tool for feature extraction; this property is in agreement with the important discovery that the human visual system indeed uses a kind of wavelet decomposition (Gabor wavelets that are based on a space-frequency decomposition) as a preprocessing tool, in particular for recognition. Furthermore, wavelets are simultaneously useful in this context for denoising. The paper first proposes a tutorial on the problem of face recognition, and then focuses on wavelet-based algorithms.

The paper by Lihua Yang gives an introduction to the Hilbert-Huang transform. It supplies its theoretical mathematical background and shows recent applications to pattern recognition. A key (but ill-posed) problem in signal processing is to define the instantaneous frequency of a signal. A classical way to define this notion is to use the Hilbert transform and the associated analytic signal; however, if applied directly to a complex signal, this method can lead to severe instabilities and ill-functionings. A fundamental advance was obtained with the introduction the Empirical Mode Decomposition, which splits the signal into simpler basic components: the Intrinsic Mode Functions; it can be efficiently used as a preprocessing, since the instantaneous frequency of each simple component can then be determined in a numerically meaningful and stable way. This paper shows recent applications of the Hilbert-Huang transform e.g. to a tsunami wave, and to pattern recognition.

Alain Damlamian, Stéphane Jaffard

Editors

May 2010

Contents

Preface

Tight Frame Based Method for High-Resolution Image Reconstruction

Jianfeng Cai[*]

Temasek Laboratories

National University of Singapore, Singapore

Email: tslcaij@nus.edu.sg

Raymond Chan[†]

Department of Mathematics

The Chinese University of Hong Kong, China

Email: rchan@math.cuhk.edu.hk

Lixin Shen[‡]

Department of Mathematics

Syracuse University, USA

Email: lshen03@syr.edu

Zuowei Shen[§]

Department of Mathematics

National University of Singapore, Singapore

Email: matzuows@nus.edu.sg

Abstract

We give a comprehensive discussion on high-resolution image reconstruction based on a tight frame. We first present the tight frame filters arising from the problem of high-resolution image reconstruction and the associated matrix representation of the filters for various boundary extensions. We then propose three algorithms for high-resolution image reconstruction using the designed tight frame filters and show analytically the properties of

[*]The research was supported by the Wavelets and Information Processing Programme under a grant from DSTA, Singapore.

[†]The research was supported in part by HKRGC Grant 400505 and CUHK DAG 2060257.

[‡]This work was supported by the US National Science Foundation under grant DMS-0712827. The work was partially done while this author was visiting the Institute for Mathematical Sciences, National University of Singapore in 2008. The visit was partially supported by the institute.

[§]The research was supported in part by Grant R-146-000-060-112 at the National University of Singapore.

these algorithms. Finally, we numerically illustrate the efficiency
of the proposed algorithms for natural images.

1 High-resolution image reconstruction model

The problem of high-resolution image reconstruction is to reconstruct a
high-resolution (HR) image from multiple, under-sampled, shifted, de-
graded and noisy frames where each frame differs from the others by some
sub-pixel shifts. The problem arises in a variety of scientific, medical,
and engineering applications. The problem of HR image reconstruction
is a hot field. In the past few years, two special issues on the topic was
published: IEEE Signal Processing Magazine (Volume 20, Issue 3, May
2003) and International Journal of Imaging Systems and Technology
(Volume 14, No. 2, 2004).

The earliest study of HR image reconstruction was motivated by the
need to improve the resolution of images from Landsat image data. In
[28], Huang and Tsay used the frequency domain approach to demon-
strate the improved reconstruction image from several down-sampled
noise-free images. Later on, Kim et al. [30] generalized this idea to
noisy and blurred images. Both methods in [28, 30] are computational
efficiency, but they are prone to model errors, and that limits their use
[1]. Statistical methods have appeared recently for super-resolution image reconstruction problems. In this direction, tools such as a maximum
a posteriori (MAP) estimator with the Huber-Markov random field prior
and a Gibbs image prior are proposed in [25, 43]. In particular, the task
of simultaneous image registration and super-resolution image reconstruction are studied in [25, 45]. Iterative spatial domain methods are
one popular class of methods for solving the problems of resolution enhancement [3, 21, 22, 23, 27, 31, 32, 36, 38, 39, 41]. The problems are
formulated as Tikhonov regularization. A great deal of work has been
devoted to the efficient calculation of the reconstruction and the estimation of the associated hyperparameters by taking advantage of the
inherent structures in the HR system matrix. Bose and Boo [3] used
a block semi-circulant matrix decomposition in order to calculate the
MAP reconstruction. Ng et al. [36] and Ng and Yip [37] proposed a
fast discrete cosine transform based approach for HR image reconstruction with Neumann boundary condition. Nguyen et al. [40, 41] also
addressed the problem of efficient calculation. The proper choice of the
regularization tuning parameter is crucial to achieving robustness in the
presence of noise and avoiding trial-and-error in the selection of an optimal tuning parameter. To this end, Bose et al. [4] used an L-curve
based approach. Nguyen et al. [41] used a generalized cross-validation

method. Molina *et al.* [33] used an expectation-maximization algorithm. Lu *et al.* [32] proposed multiparameter regularization methods which introduce different regularization parameters for different frequency bands of the regularization operator.

Low-resolution images can be viewed as outputs of the original high-resolution image passing through a low-pass filter followed by a decimation process. This viewpoint suggests that a framework of multiresolution analysis can be naturally adopted to produce an HR image from a set of low-resolution images of the same scene with sub-pixel shifts. In this fashion, a series of work has been done recently, see, e.g., [9, 10, 11, 12, 13]. Extension of these work will be discussed in the paper.

Here we present a mathematical model proposed by Bose and Boo in [3] for high-resolution image reconstruction. Consider $K \times K$ sub-window-shifted low-resolution images in which each image has $N_1 \times N_2$ interrogation windows and the size of each interrogation window is $T_1 \times T_2$. Here, $K \times K$ denotes K shifts in both the vertical and horizontal directions. The goal is to reconstruct a much higher resolution image with $M_1 \times M_2$ sub-windows, where $M_1 = K \times N_1$ and $M_2 = K \times N_2$.

In order to have enough information to resolve the high-resolution image, it is assumed that there are sub-window shifts between the low-resolution images. For a low-resolution image denoted by (k_1, k_2), where $0 \leq k_1, k_2 < K$ with $(k_1, k_2) \neq (0, 0)$, its vertical and horizontal shifts $d^x_{k_1,k_2}$ and $d^y_{k_1,k_2}$ with respect to the $(0,0)$th reference low-resolution image are given by $d^x_{k_1,k_2} = \left(k_1 + \epsilon^x_{k_1,k_2}\right) \frac{T_1}{K}$ and $d^y_{k_1,k_2} = \left(k_2 + \epsilon^y_{k_1,k_2}\right) \frac{T_2}{K}$. Here $\epsilon^x_{k_1,k_2}$ and $\epsilon^y_{k_1,k_2}$ are the vertical and horizontal *shift errors* respectively. We assume that $|\epsilon^x_{k_1,k_2}| < \frac{1}{2}$ and $|\epsilon^y_{k_1,k_2}| < \frac{1}{2}$. Figure 1.1 shows the example of 2×2 shifted low-resolution images.

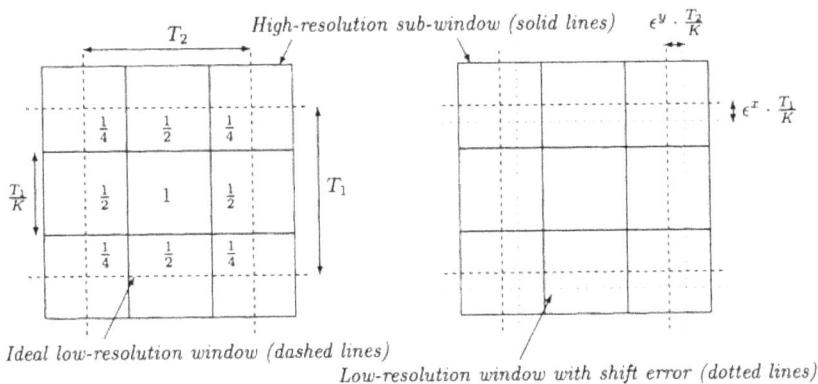

Figure 1.1   Windows without and with shift error when $K = 2$ (left and right respectively).

For a low-resolution image (k_1, k_2), the average quantity at its (n_1, n_2)th interrogation window is modelled by:

$$g_{k_1,k_2}[n_1, n_2] = \frac{1}{T_1 T_2} \int_{A_{k_1,k_2:n_1,n_2}} f(x, y) \, dx \, dy + \eta_{k_1,k_2}[n_1, n_2], \quad (1.1)$$

where the interrogation window in the low-resolution image is

$$A_{k_1,k_2:n_1,n_2} = \left[T_1 \left(n_1 - \frac{1}{2} \right) + d_{k_1,k_2}^x, T_1 \left(n_1 + \frac{1}{2} \right) + d_{k_1,k_2}^x \right]$$
$$\times \left[T_2 \left(n_2 - \frac{1}{2} \right) + d_{k_1,k_2}^y, T_2 \left(n_2 + \frac{1}{2} \right) + d_{k_1,k_2}^y \right].$$

Here (n_1, n_2) indicates an interrogation window in the low-resolution image (k_1, k_2) (where $0 \le n_1 < N_1$ and $0 \le n_2 < N_2$) and $\eta_{k_1,k_2}[n_1, n_2]$ is the noise (refer to [3]). We interlace all the sub-window-shifted low-resolution images g_{k_1,k_2} to form an $M_1 \times M_2$ image g by assigning

$$g[Kn_1 + k_1, Kn_2 + k_2] = g_{k_1,k_2}[n_1, n_2].$$

The pseudo high-resolution image g is called the *observed high-resolution image*.

The integral values on the sub-window of the high-resolution image is approximated by

$$f[i, j] = \frac{K^2}{T_1 T_2} \int_{A_{i,j}} f(x, y) \, dx \, dy, \quad 0 \le i < M_1, 0 \le j < M_2, \quad (1.2)$$

which is the average quantity inside the (i, j)th high-resolution sub-window:

$$A_{i,j} = \left[i\frac{T_1}{K}, (i + 1)\frac{T_1}{K} \right] \times \left[j\frac{T_2}{K}, (j + 1)\frac{T_2}{K} \right], \quad 0 \le i < M_1, 0 \le j < M_2.$$
$$(1.3)$$

To obtain the true high-resolution image f from the observed high-resolution image g, one will have to solve (1.1) for f. By discretizing (1.1) and (1.2) using the rectangular quadrature rule, we have

$$g_{k_1,k_2}[n_1, n_2] = \sum_{p,q=0}^{K} W[p, q]f[Kn_1 + k_1 + p, Kn_2 + k_2 + q] + \eta_{k_1,k_2}[n_1, n_2],$$
$$(1.4)$$

where the weighting matrix W for discretizing the integral equation (1.1) in the case without shift error is

$$
W = \frac{1}{K^2}
\begin{bmatrix}
\frac{1}{4} & \frac{1}{2} & \cdots & \frac{1}{2} & \frac{1}{4} \\[4pt]
\frac{1}{2} & 1 & \cdots & 1 & \frac{1}{2} \\[4pt]
\vdots & \vdots & \ddots & \vdots & \vdots \\[4pt]
\frac{1}{2} & 1 & \cdots & 1 & \frac{1}{2} \\[4pt]
\frac{1}{4} & \frac{1}{2} & \cdots & \frac{1}{2} & \frac{1}{4}
\end{bmatrix},
\tag{1.5}
$$

which is assigned for associated sub-windows of the high-resolution image. Equation (1.4) is a system of linear equations relating the unknown values $f[i, j]$ to the given observed high-resolution image values $g[i, j]$.

For simplifying the exposition, f and g will be considered as the column vectors formed by $f[i, j]$ and $g[i, j]$. This linear system corresponding to (1.4) for high-resolution image reconstruction is reduced to

$$
Hf + \eta = g,
\tag{1.6}
$$

where the blurring matrix H, which is formulated from (1.4), varies under different boundary conditions and η is the noise vector. For the case without shift error, the blurring matrix H is given by

$$
H = \frac{1}{K^4}
\begin{bmatrix}
\ddots & \ddots & \ddots & \ddots & \ddots \\
\frac{1}{2} & 1 & \cdots & 1 & \frac{1}{2} \\
& \ddots & \ddots & \ddots & \ddots & \ddots
\end{bmatrix}
\otimes
\begin{bmatrix}
\ddots & \ddots & \ddots & \ddots & \ddots \\
\frac{1}{2} & 1 & \cdots & 1 & \frac{1}{2} \\
& \ddots & \ddots & \ddots & \ddots & \ddots
\end{bmatrix},
$$

where the Kronecker operator $\otimes$ is defined by $A \otimes B = [a_{ij}B]$ with $A = [a_{ij}]$. The key problem is to recover the true high-resolution image f from the observed high-resolution image g by solving (1.6).

If the low-resolution images are shifted by exactly half of the window, then the problem reduces to solving a spatially invariant linear system. Depending on the boundary conditions we impose on the images, the coefficient matrix H is either Topelitz or Toeplitz-like. The model was then solved in [3, 36] using preconditioned conjugate gradient method.

We next discuss in details several approaches that will use the tight frame for solving the system (1.4) or (1.6). The performance of these methods will be examined in numerical simulations. In the next section, we will give a brief review on the frame theory. In particular, we will present the tight frame system with (1.5) as its low-pass filter.

The outline of this paper is as follows. In Section 2, we give a brief review on tight frames with an emphasis on the unitary extension principle. Section 3 contains four main parts. The first part presents the tight frames arising from the problem of HR image reconstruction. The matrix representations of the tight frame filters associated with the HR image reconstruction are given, by imposing the periodic and symmetric boundary conditions, in the second and third parts, respectively. It follows by showing the multi-level framelet decomposition and reconstruction in the last part. We propose three framelet-based algorithms to tackle the problem of HR image reconstruction in Section 4. In particular, we give a complete analysis for Algorithm I in Section 5. Numerical experiments for all three algorithms are presented in Section 6.

2   Preliminaries on tight framelets

The notion of frame was first introduced by Duffin and Schaeffer [20] in 1952. A countable system $X \subset L^2(\mathbb{R})$ is called a *frame* of $L^2(\mathbb{R})$ if

$$\alpha \|f\|_2^2 \le \sum_{h \in X} |\langle f, h \rangle|^2 \le \beta \|f\|_2^2, \qquad (2.1)$$

where the constants α and β, $0 < \alpha \le \beta < \infty$, are lower and upper bounds of the frame system X. The notation $\langle \cdot, \cdot \rangle$ and $\| \cdot \|_2 = \langle \cdot, \cdot \rangle^{1/2}$ are the inner product and norm of $L^2(\mathbb{R})$. When $\alpha = \beta(= 1)$, the frame system X is called a *tight frame*. In what follows, our discussion is concentrated on the tight frame.

Two operators, namely analysis operator and synthesis operator, are associated with the tight frame. The analysis operator of the frame is defined as

$$\mathcal{F} : L^2(\mathbb{R}) \longrightarrow \ell^2$$

with

$$\mathcal{F}(f) = \{\langle f, h \rangle\}_{h \in X}.$$

Its adjoint operator $\mathcal{F}^*$, called the synthesis operator, is defined as

$$\mathcal{F}^* : \ell^2 \longrightarrow L^2(\mathbb{R})$$

with

$$\mathcal{F}^*(c) = \sum_{h \in X} c_h h, \quad c = \{c_h\}_{h \in X}.$$

Hence, X is a tight frame if and only if $\mathcal{F}^* \mathcal{F} = \mathcal{I}$. This is true if

$$f = \sum_{h \in X} \langle f, h \rangle h, \quad \forall f \in L^2(\mathbb{R}), \qquad (2.2)$$

which is equivalent to

$$\|f\|_2^2 = \sum_{h \in X} |\langle f, h \rangle|^2, \quad \forall f \in L^2(\mathbb{R}). \tag{2.3}$$

Equation (2.2) is the perfect reconstruction formula of the tight frame. Identities (2.2) and (2.3) hold for an arbitrary orthonormal basis of $L^2(\mathbb{R})$. In this sense, an orthonormal basis is a tight frame, and a tight frame is a generalization of orthonormal basis. But tight frames sacrifice the orthonormality and the linear independence of the system in order to get more flexibility. Therefore tight frames can be redundant.

For a tight frame system X, we have

$$\sum_{h \in X} |\langle f, h \rangle|^2 \leq \sum_{h \in X} |c_h|^2$$

for all possible representation of $f = \sum_{h \in X} c_h h$, $\{c_h\} \in \ell^2$. In other words, the sequence $\mathcal{F}(f)$ obtained by the analysis operator $\mathcal{F}$ has the smallest ℓ^2 norm among all sequences $\{c_h\} \in \ell^2$ satisfying $f = \sum_{h \in X} c_h h$.

If $X(\Psi)$ is the collection of the dilations and the shifts of a finite set $\Psi \subset L^2(\mathbb{R})$, i.e.,

$$X(\Psi) = \{K^{i/2} \psi(K^i x - j) : \psi \in \Psi, i, j \in \mathbb{Z}\},$$

then $X(\Psi)$ is called a *wavelet* (or *affine*) *system of dilation* K. In this case the elements in Ψ are called the *generators*. When $X(\Psi)$ is a tight frame for $L^2(\mathbb{R})$, then $\psi \in \Psi$ are called (*tight*) *framelets*.

A normal framelet construction starts with a refinable function. A compactly supported function $\phi \in L^2(\mathbb{R})$ is *refinable* (a scaling function) with a refinement mask τ_ϕ if it satisfies

$$\widehat{\phi}(K \cdot) = \tau_\phi \widehat{\phi}.$$

Here $\widehat{\phi}$ is the Fourier transform of ϕ, and τ_ϕ is a trigonometric polynomial with $\tau_\phi(0) = 1$, i.e., a refinement mask of a refinable function must be a lowpass filter. One can define a multiresolution analysis from a given refinable function, details about that is omitted here, but can be found, for instance, in [19, 29].

For a given compactly supported refinable function, the construction of tight framelet systems is to find a finite set Ψ that can be represented in the Fourier domain as

$$\widehat{\psi}(K \cdot) = \tau_\psi \widehat{\phi}$$

for some 2π-periodic τ_ψ. The unitary extension principle (UEP) of [42] says that the wavelet system $X(\Psi)$ generated by a finite set Ψ forms a tight frame in $L^2(\mathbb{R})$ provided that the masks τ_ϕ and $\{\tau_\psi\}_{\psi\in\Psi}$ satisfy:

$$\overline{\tau_\phi(\omega)}\tau_\phi\left(\omega+\frac{2\gamma\pi}{K}\right)+\sum_{\psi\in\Psi}\overline{\tau_\psi(\omega)}\tau_\psi\left(\omega+\frac{2\gamma\pi}{K}\right)=\delta_{\gamma,0},$$

$$\gamma=0,1,\cdots,K-1 \qquad (2.4)$$

for almost all ω in $\mathbb{R}$. Practically, we require all masks to be trigono-metric polynomials. Thus, (2.4) together with the fact that $\tau_\phi(0)=1$ imply that $\tau_\psi(0)=0$ for all $\psi\in\Psi$. Hence, $\{\tau_\psi\}_{\psi\in\Psi}$ must correspond to high-pass filters. The sequences of Fourier coefficients of τ_ψ, as well as τ_ψ itself, are called *framelet masks*. The construction of framelets Ψ essentially is to design, for a given refinement mask τ_ϕ, framelet masks $\{\tau_\psi\}_{\psi\in\Psi}$ such that (2.4) holds. A more general principle of construction tight framelets, the oblique extension principle, was developed recently in [14, 17].

In the next section, we will use the EUP to construct a framelet system arising from the problem of HR image reconstruction.

3 Tight frame system arising from high-resolution image reconstruction

3.1   Filter design

The low-pass filter (1.5) for high-resolution image reconstruction is sep-arable and can be written as follows

$$W=h_0^T h_0,$$

where

$$h_0=\frac{1}{K}\left[\frac{1}{2},1,\cdots,1,\frac{1}{2}\right].$$

Hence, to design a tight frame system with W as its low-pass filter, we just need to construct a tight frame system with h_0 as its low-pass filter. By virtue of the Fourier series of h_0, define

$$\widehat{\phi}(\omega):=\prod_{j=1}^{\infty}\widehat{h}_0(K^{-j}\omega), \qquad (3.1)$$

where

$$\widehat{h}_0(\omega)=\frac{1}{2K}+\frac{1}{K}\left(\sum_{k=1}^{K-1}\exp(-ik\omega)\right)+\frac{1}{2K}\exp(-iK\omega).$$

It was shown in [44] that ϕ is in $L^2(\mathbb{R})$ and Hölder continuous with Hölder exponent $\ln 2/\ln K$.

For any integer $L \geq 2$, define

$$m_{L,p} := \frac{\sqrt{2}}{L}\left[\cos\left(\frac{p\pi}{2L}\right), \cos\left(\frac{3p\pi}{2L}\right), \cdots, \cos\left(\frac{(2L-1)p\pi}{2L}\right)\right],$$

and their Fourier series

$$\widehat{m}_{L,p}(\omega) = \frac{\sqrt{2}}{L}\sum_{\ell=1}^{L}\cos\left(\frac{(2\ell-1)p\pi}{2L}\right)\exp(-i\ell\omega),$$

for $p = 1, \cdots, L-1$. We further define, for any integer $K \geq 2$,

$$\widehat{h}_{2p+q}(\omega) := \widehat{m}_{2,q}(\omega)\widehat{m}_{K,p}(\omega), \tag{3.2}$$

where $p = 1, \cdots, K-1$, $q \in \{0,1\}$. We can easily check that

$$\sum_{q=0}^{1}\sum_{p=0}^{K-1}\widehat{h}_{2p+q}(\omega)\overline{\widehat{h}_{2p+q}(\omega + \frac{2\pi\ell}{K})} = \delta_{\ell,0}, \quad \ell = 0, 1, \cdots, K-1. \tag{3.3}$$

The EUP of [42] yields that the functions

$$\Psi = \{\psi_{2p+q} : 0 \leq p \leq K-1, \quad q = 0, 1, \quad (p,q) \neq (0,0)\}$$

defined by

$$\widehat{\psi}_{2p+q}(\omega) = \widehat{h}_{2p+q}\left(\frac{\omega}{K}\right)\widehat{\phi}\left(\frac{\omega}{K}\right)$$

are tight framelets. Furthermore,

$$X(\Psi) = \left\{K^{k/2}\psi_{2p+q}(K^k \cdot -j) : 0 \leq p \leq K-1,\right.$$
$$\left. q = 0, 1, (p,q) \neq (0,0); k, j \in \mathbb{Z}\right\}$$

is a tight frame system of $L^2(\mathbb{R})$.

In the following discussion, we always assume that the indexes of all filters h_ℓ, run from $-K/2$ to $K/2$ for even number K and $-(K+1)/2$ to $(K-1)/2$ for odd number. We are interested in the matrix representation of the identity

$$\sum_{q=0}^{1}\sum_{p=0}^{K-1}|\widehat{h}_{2p+q}(\omega)|^2 = 0 \tag{3.4}$$

for filters given by (3.2). In image processing, periodic and symmetric boundary conditions are usually imposed to give matrix representation of (3.4). In the following subsections, we will give the corresponding representations for both boundary conditions.

3.2 Matrix representation of filters with periodic boundary conditions

For simplicity, we are not going to write the matrix forms of the filters given by (3.2) for a general integer K. Instead, we give these matrices for the filters with $K = 2$ and $K = 3$ only. From there, one can easily give the matrix representation for filters associated with any integer K.

Example 1. *For $K = 2$, we have, from (3.2), the low-pass filter $h_0 = \frac{1}{4}[1,2,1]$ and three high-pass filters $h_1 = \frac{1}{4}[1,0,-1]$, $h_2 = \frac{1}{4}[1,0,-1]$, and $h_3 = \frac{1}{4}[1,-2,1]$, respectively. The corresponding matrix representation under the periodic boundary condition for filters h_0, h_1, h_2, and h_3 are circulant matrices H_0, H_1, H_2, and H_3, respectively, with their first rows being the following*

$$\left[\frac{1}{2},\frac{1}{4},0,\cdots,0,\frac{1}{4}\right], \quad \left[0,-\frac{1}{4},0,\cdots,0,\frac{1}{4}\right],$$

$$\left[0,-\frac{1}{4},0,\cdots,0,\frac{1}{4}\right], \quad \left[-\frac{1}{2},\frac{1}{4},0,\cdots,0,\frac{1}{4}\right].$$

We can check that

$$H_0^T H_0 + H_1^T H_1 + H_2^T H_2 + H_3^T H_3 = I.$$

We remark that $h_1 = h_2$ in above tight frame filters. We can merge h_1 and h_2 together and deduce a new tight frame system with the low-pass filter $\frac{1}{4}[1,2,1]$ and two high-pass filters $\frac{\sqrt{2}}{4}[1,0,-1]$ and $\frac{1}{4}[1,-2,1]$. A similar situation happens in the next example.

Example 2. *For $K = 3$, we have the low-pass filter $h_0 = \frac{1}{6}[1,2,2,1]$ and five high-pass filters $h_1 = \frac{1}{6}[1,0,0,-1]$, $h_2 = \frac{\sqrt{6}}{12}[1,1,-1,-1]$, $h_3 = \frac{\sqrt{6}}{12}[1,-1,-1,1]$, $h_4 = \frac{\sqrt{2}}{12}[1,-1,-1,1]$, and $h_5 = \frac{\sqrt{2}}{12}[1,-3,3,-1]$. The corresponding matrix representation under the periodic boundary condition for filters $h_0,h_1,\cdots,h_5$ are circulant matrices $H_0,H_1,\cdots,H_5$, respectively, with their first rows being the following*

$$\frac{1}{6}[2,1,0,\cdots,0,1,2], \quad \frac{1}{6}[0,-1,0,\cdots,0,1,0],$$

$$\frac{\sqrt{6}}{12}[-1,-1,0,\cdots,0,1,1], \quad \frac{\sqrt{6}}{12}[-1,1,0,\cdots,0,1,-1],$$

$$\frac{\sqrt{2}}{12}[-1,1,0,\cdots,0,1,-1], \quad \frac{\sqrt{2}}{12}[3,-1,0,\cdots,0,1,-3].$$

Again, it can be easily checked that

$$H_0^T H_0 + H_1^T H_1 + H_2^T H_2 + H_3^T H_3 + H_4^T H_4 + H_5^T H_5 = I.$$

3.3 Matrix representation of filters with symmetric boundary conditions

A filter h is said to be symmetric (or antisymmetric) with symmetric center $\frac{N}{2}$ for some integer N if

$$h[N-k] = h[k], \quad k \in \mathbb{Z} \quad (\text{or} \quad h[N-k] = -h[k], \quad k \in \mathbb{Z}).$$

We denote $c(h) = \frac{N}{2}$ the symmetric center of the symmetric (or antisymmetric) filter h. With this definition, we immediately have the following result.

Proposition 1. *Let h_ℓ be the filters constructed by (3.2). Then $c(h_\ell) = 0$ for even K and $c(h_\ell) = \frac{1}{2}$ for odd K.*

For any infinite signal $u = (\cdots, u(-1), u(0), u(1), \cdots)^t$, where $u(k) \in \mathbb{R}$, $k \in \mathbb{Z}$, if there exists an integer n such that $u(n-k) = u(k)$ (or $u(n-k) = -u(k)$), for all $k \in \mathbb{Z}$, then we say that u is symmetric (or antisymmetric) with the center $c(u) = \frac{n}{2}$. Accordingly, if $c(u) = \frac{n}{2}$ is an integer, it is referred to as whole-sample symmetric (or whole-sample antisymmetric) and denoted by WS (or WA); if $c(v) = \frac{n}{2}$ is not an integer, it is referred to as half-sample symmetric (or half-sample antisymmetric) and denoted by HS (or HA). If u has two centers c_1 and c_2 with $c_1 < c_2$, we denote it by $c(u) = (c_1, c_2)$.

For a finite-length signal

$$u = (u(0), \cdots, u(N-1))^t,$$

the infinite signal $v = (\cdots, v(-1), v(0), v(1), \cdots)^t$ is an extended signal of u if $v(k) = u(k)$, for $0 \le k \le N-1$. We define a restriction operator P_N as follows:

$$P_N : \mathbb{R}^\infty \to \mathbb{R}^N; \quad (\cdots, x(-1), x(0), \cdots, x(N-1), x(N), \cdots)^t$$
$$\mapsto (x(0), \cdots, x(N-1))^t.$$

Hence, if v is an extended signal of a finite length signal u of length N, then $u = P_N(v)$.

There are many ways to extend a signal into another signal with infinite length. We are mostly interested in a symmetric extension method since all filters h_ℓ are symmetric or antisymmetric. By doing so, our goal is to construct matrices H_ℓ and $\underline{H}_\ell$ associated with filters h_ℓ and $\underline{h}_\ell$ ($\underline{h}_\ell[k] = h_\ell[-k]$) of size $N \times N$ such that

$$\sum_\ell \underline{H}_\ell H_\ell = I. \tag{3.5}$$

Clearly, the properties of extension methods are reflected in the structures of the matrices H_ℓ and $\underline{H}_\ell$.

To develop the matrix representation of the filters in (3.2), we therefore restrict ourselves to the following extension methods:

- The whole-sample symmetric extension (WSWS) $E_s^{(1,1)}$:

$$E_s^{(1,1)}(u) = (\cdots, u(2), u(1), u(0), u(1), \cdots,$$
$$u(N-2), u(N-1), u(N-2), u(N-3), \cdots)^t.$$

- The whole-half-sample extension (WSHS) $E_s^{(1,2)}$:

$$E_s^{(1,2)}(u) = (\cdots, u(2), u(1), u(0), u(1), \cdots, u(N-2), u(N-1),$$
$$u(N-1), u(N-2), u(N-3), \cdots)^t.$$

- The half-whole-sample extension (HSWS) $E_s^{(2,1)}$:

$$E_s^{(2,1)}(u) = (\cdots, u(1), u(0), u(0), u(1), \cdots, u(N-2), u(N-1),$$
$$u(N-2), u(N-3), \cdots)^t.$$

- The whole-sample antisymmetric extension (WAWA) $E_a^{(1,1)}$:

$$E_a^{(1,1)}(u) = (\cdots, -u(2), -u(1), u(0), u(1), \cdots, u(N-2),$$
$$u(N-1), -u(N-2), -u(N-3), \cdots)^t.$$

- The whole-half-sample antisymmetric extension (WAHA) $E_a^{(1,2)}$:

$$E_a^{(1,2)}(u) = (\cdots, -u(2), -u(1), u(0), u(1), \cdots, u(N-2),$$
$$u(N-1), -u(N-1), -u(N-2), \cdots)^t.$$

- The half-whole-sample antisymmetric extension (HAWA) $E_a^{(2,1)}$:

$$E_a^{(2,1)}(u) = (\cdots, -u(1), -u(0), u(0), u(1), \cdots, u(N-2),$$
$$u(N-1), -u(N-2), -u(N-3), \cdots)^t.$$

Clearly, $c(E_s^{(1,1)}(u)) = (0, N-1)$, $c(E_s^{(1,2)}(u)) = (0, N-\frac{1}{2})$, $c(E_s^{(2,1)}(u))$ $= (-\frac{1}{2}, N-1)$, $c(E_a^{(1,1)}(u)) = (0, N-1)$, $c(E_a^{(1,2)}(u)) = (0, N-\frac{1}{2})$, and $c(E_a^{(2,1)}(u)) = (-\frac{1}{2}, N-1)$.

Proposition 2. *Let $u = (u(0), \cdots, u(N-1))^t$ be a signal of length N. Then*

1. *$h * E_s^{(1,1)}(u)$ is symmetric with centers 0 and $N-1$ if h is a symmetric, odd length filter with $c(h) = 0$;*

2. *$h * E_s^{(1,1)}(u)$ is antisymmetric with centers 0 and $N-1$ if h is an antisymmetric, odd length filter with $c(h) = 0$;*

3. $h * E_s^{(2,2)}(u)$ is symmetric with centers $-\frac{1}{2}$ and $N - \frac{1}{2}$ if h is a symmetric, odd length filter with $c(h) = 0$;

4. $h * E_s^{(2,2)}(u)$ is antisymmetric with centers $-\frac{1}{2}$ and $N - \frac{1}{2}$ if h is an antisymmetric, odd length filter with $c(h) = 0$;

5. $h * E_a^{(1,1)}(u)$ is symmetric with centers 0 and $N - 1$ if h is an antisymmetric, odd length filter with $c(h) = 0$;

6. $h * E_s^{(1,2)}(u)$ is symmetric with centers $-\frac{1}{2}$ and $N - 1$ if h is a symmetric, even length filter with $c(h) = \frac{1}{2}$;

7. $h * E_s^{(1,2)}(u)$ is antisymmetric with centers $-\frac{1}{2}$ and $N - 1$ if h is an antisymmetric, even length filter with $c(h) = \frac{1}{2}$;

8. $h * E_s^{(2,1)}(u)$ is symmetric with centers 0 and $N - \frac{1}{2}$ if h is a symmetric, even length filter with $c(h) = -\frac{1}{2}$;

9. $h * E_a^{(2,1)}(u)$ is symmetric with centers 0 and $N - \frac{1}{2}$ if h is an antisymmetric, even length filter with $c(h) = -\frac{1}{2}$.

The proof of Proposition 2 is straightforward and will be omitted here.

Now, we will show how to form matrices H_ℓ and $\underline{H}_\ell$ in (3.5). Without loss of generality, we show these matrices for the filters with $K = 2$ and $K = 3$ only. In a similar fashion, one can give the matrix representation for filters associated with any integer K.

Example 3. For $K = 2$, we have the low-pass filter $h_0 = \frac{1}{4}[1, 2, 1]$ and three high-pass filters $h_1 = \frac{1}{4}[1, 0, -1]$, $h_2 = \frac{1}{4}[1, 0, -1]$, and $h_3 = \frac{1}{4}[1, -2, 1]$, respectively. If we apply the whole-point symmetric extension for the input signal, then

$$
H_0 = \frac{1}{4}\begin{bmatrix} 2 & 2 & & & \\ 1 & 2 & 1 & & \\ & \ddots & \ddots & \ddots & \\ & & 1 & 2 & 1 \\ & & & 2 & 2 \end{bmatrix} \quad H_1 = \frac{1}{4}\begin{bmatrix} 0 & 0 & & & \\ 1 & 0 & -1 & & \\ & \ddots & \ddots & \ddots & \\ & & 1 & 0 & -1 \\ & & & 0 & 0 \end{bmatrix},
$$

$$
H_2 = H_1, \quad H_3 = \frac{1}{4}\begin{bmatrix} -2 & 2 & & & \\ 1 & -2 & 1 & & \\ & \ddots & \ddots & \ddots & \\ & & 1 & -2 & 1 \\ & & & 2 & -2 \end{bmatrix}.
$$

By Items 1 and 2 of Proposition 2, matrices $\underline{H}_\ell$ are

$$\underline{H}_0 = H_0, \quad \underline{H}_1 = \frac{1}{4}\begin{bmatrix} 0 & 2 & & & \\ -1 & 0 & 1 & & \\ & \ddots & \ddots & \ddots & \\ & & -1 & 0 & 1 \\ & & & -2 & 0 \end{bmatrix}, \quad \underline{H}_2 = \underline{H}_1, \quad \underline{H}_3 = H_3.$$

If we apply the half-point extension for the input signal, then

$$H_0 = \frac{1}{4}\begin{bmatrix} 3 & 1 & & & \\ 1 & 2 & 1 & & \\ & \ddots & \ddots & \ddots & \\ & & 1 & 2 & 1 \\ & & & 1 & 3 \end{bmatrix}, \quad H_1 = \frac{1}{4}\begin{bmatrix} 1 & -1 & & & \\ 1 & 0 & -1 & & \\ & \ddots & \ddots & \ddots & \\ & & 1 & 0 & -1 \\ & & & 1 & -1 \end{bmatrix},$$

$$H_2 = H_1, \quad H_3 = \frac{1}{4}\begin{bmatrix} -1 & 1 & & & \\ 1 & -2 & 1 & & \\ & \ddots & \ddots & \ddots & \\ & & 1 & -2 & 1 \\ & & & 1 & -1 \end{bmatrix}.$$

By Items 3 and 4 of Proposition 2, matrices $\underline{H}_\ell$ are

$$\underline{H}_0 = H_0, \quad \underline{H}_1 = \frac{1}{4}\begin{bmatrix} 1 & 1 & & & \\ -1 & 0 & 1 & & \\ & \ddots & \ddots & \ddots & \\ & & -1 & 0 & 1 \\ & & & -1 & -1 \end{bmatrix}, \quad \underline{H}_2 = \underline{H}_1, \quad \underline{H}_3 = \underline{H}_0.$$

In both the whole-point and half-point extensions, the perfect reconstruction (3.5) is satisfied. Furthermore, for the half-point extension for the input signal, we have

$$H_0^T H_0 + H_1^T H_1 + H_2^T H_2 + H_3^T H_3 = I.$$

Example 4. For $K = 3$, we have the low-pass filter $h_0 = \frac{1}{6}[1,2,2,1]$ and five high-pass filters $h_1 = \frac{1}{6}[1,0,0,-1]$, $h_2 = \frac{\sqrt{6}}{12}[1,1,-1,-1]$, $h_3 = \frac{\sqrt{6}}{12}[1,-1,-1,1]$, $h_4 = \frac{\sqrt{2}}{12}[1,-1,-1,1]$, and $h_5 = \frac{\sqrt{2}}{12}[1,-3,3,-1]$. Indices of these filters are from -2 to 1. If we apply the half-point extension at the left end and the whole-point at the right end for the input signal, *we have*

$$H_0 = \frac{1}{6}\begin{bmatrix} 4 & 2 & & & & \\ 3 & 2 & 1 & & & \\ 1 & 2 & 2 & 1 & & \\ & \ddots & \ddots & \ddots & \ddots & \\ & & 1 & 2 & 2 & 1 \\ & & & 1 & 3 & 2 \end{bmatrix}, \quad H_1 = \frac{1}{6}\begin{bmatrix} 0 & 0 & & & & \\ 1 & 0 & -1 & & & \\ 1 & 0 & 0 & -1 & & \\ & \ddots & \ddots & \ddots & \ddots & \\ & & 1 & 0 & 0 & -1 \\ & & & 1 & -1 & 0 \end{bmatrix},$$

$$H_2 = \frac{\sqrt{6}}{12}\begin{bmatrix} 0 & 0 & & & & \\ 2 & -1 & 1 & & & \\ 1 & 1 & -1 & -1 & & \\ & \ddots & \ddots & \ddots & \ddots & \\ & & 1 & 1 & -1 & -1 \\ & & & 1 & 0 & -1 \end{bmatrix}, \quad H_3 = \frac{\sqrt{6}}{12}\begin{bmatrix} -2 & 2 & & & & \\ 0 & -1 & 1 & & & \\ 1 & -1 & -1 & 1 & & \\ & \ddots & \ddots & \ddots & \ddots & \\ & & 1 & -1 & -1 & 1 \\ & & & 1 & 0 & -1 \end{bmatrix},$$

$$H_4 = \frac{\sqrt{3}}{3}H_3, \quad H_5 = \frac{\sqrt{2}}{12}\begin{bmatrix} 0 & 0 & & & & \\ -2 & 3 & -1 & & & \\ 1 & -3 & 3 & -1 & & \\ & \ddots & \ddots & \ddots & \ddots & \\ & & 1 & -3 & 3 & -1 \\ & & & 1 & -4 & 3 \end{bmatrix},$$

$$\underline{H}_0 = \frac{1}{6}\begin{bmatrix} 2 & 3 & 1 & & & \\ 1 & 2 & 2 & 1 & & \\ & \ddots & \ddots & \ddots & \ddots & \\ & & 1 & 2 & 2 & 1 \\ & & & 1 & 2 & 3 \\ & & & & 2 & 4 \end{bmatrix}, \quad \underline{H}_1 = \frac{1}{6}\begin{bmatrix} 0 & 1 & 1 & & & \\ -1 & 0 & 0 & 1 & & \\ & \ddots & \ddots & \ddots & \ddots & \\ & & -1 & 0 & 0 & 1 \\ & & & -1 & 0 & -1 \\ & & & & -2 & 0 \end{bmatrix},$$

$$\underline{H}_2 = \frac{\sqrt{6}}{12}\begin{bmatrix} -1 & 2 & 1 & & & \\ -1 & -1 & 1 & 1 & & \\ & \ddots & \ddots & \ddots & \ddots & \\ & & -1 & -1 & 1 & 1 \\ & & & -1 & -1 & 0 \\ & & & & -2 & -2 \end{bmatrix}, \quad \underline{H}_3 = \frac{\sqrt{6}}{12}\begin{bmatrix} -1 & 0 & 1 & & & \\ 1 & -1 & -1 & 1 & & \\ & \ddots & \ddots & \ddots & \ddots & \\ & & 1 & -1 & -1 & 1 \\ & & & 1 & -1 & 0 \\ & & & & 2 & -2 \end{bmatrix},$$

$$\underline{H}_4 = \frac{\sqrt{3}}{3}\underline{H}_3, \quad \underline{H}_5 = \frac{\sqrt{2}}{12}\begin{bmatrix} 3 & -2 & 1 & & & \\ -1 & 3 & -3 & 1 & & \\ & \ddots & \ddots & \ddots & \ddots & \\ & & -1 & 3 & -3 & 1 \\ & & & -1 & 3 & -4 \\ & & & & -2 & 6 \end{bmatrix}.$$

3.4   Multi-level framelet decomposition and reconstruction

To analyze the given signal via a tight frame, one needs to decompose the signal in different levels in the transform domain. This process can be accomplished through a matrix A which is associated with the underlying framelet system. Since Ron and Shen's piecewise linear tight frame [42] will be used in our algorithms developed in the next section, we just give the matrix A for this particular tight frame.

The piecewise linear tight frame is generated by the following low-pass h_0, bandpass h_1, and high-pass h_2 filters:

$$h_0 = \frac{1}{4}[1,2,1], \quad h_1 = \frac{\sqrt{2}}{4}[1,0,-1], \quad \text{and} \quad h_2 = \frac{1}{4}[1,-2,1].$$

The scaling function ϕ and the wavelets ψ_1 and ψ_2, associated with h_0, h_1, and h_2, respectively, are given in the Fourier domain by

$$\widehat{\phi}(\omega) = \frac{\sin^2(\omega/2)}{(\omega/2)^2}, \quad \widehat{\psi}_1(\omega) = i\sqrt{2}\frac{\cos(\omega/4)\sin^3(\omega/4)}{(\omega/4)^2},$$

$$\text{and} \quad \widehat{\psi}_2(\omega) = -\sqrt{2}\frac{\sin^4(\omega/4)}{(\omega/4)^2}.$$

For any non-negative integer ℓ, we define

$$h_{0,\ell} := \frac{1}{4}[1,\underbrace{0,\cdots,0}_{2^{\ell-1}-1},2,\underbrace{0,\cdots,0}_{2^{\ell-1}-1},1],$$

$$h_{1,\ell} := \frac{\sqrt{2}}{4}[1,\underbrace{0,\cdots,0}_{2^{\ell-1}-1},0,0,\underbrace{0,\cdots,0}_{2^{\ell-1}-1},-1],$$

$$h_{2,\ell} := \frac{1}{4}[1,\underbrace{0,\cdots,0}_{2^{\ell-1}-1},-2,\underbrace{0,\cdots,0}_{2^{\ell-1}-1},1].$$

Following Example 3 with a half-point symmetric extension, we define matrices $H_0^{(\ell)}$, $H_1^{(\ell)}$, and $H_2^{(\ell)}$ associated with filters $h_{0,\ell}$, $h_{1,\ell}$, and $h_{2,\ell}$, respectively. Further, we have

$$(H_0^{(\ell)})^T H_0^{(\ell)} + (H_1^{(\ell)})^T H_1^{(\ell)} + (H_2^{(\ell)})^T H_2^{(\ell)} = I.$$

Now, the matrix A corresponding to the L-level framelet decomposition is

$$A := \begin{bmatrix} \prod_{\ell=0}^{L-1} H_0^{(L-\ell)} \\ H_1^{(L)} \prod_{\ell=1}^{L-1} H_0^{(L-\ell)} \\ H_2^{(L)} \prod_{\ell=1}^{L-1} H_0^{(L-\ell)} \\ \vdots \\ H_1^{(1)} \\ H_2^{(1)} \end{bmatrix}. \tag{3.6}$$

Obviously, the L-level perfect reconstruction formula is

$$A^T A = I.$$

We remark that a different matrix A is used in our Algorithm III proposed in the next section.

4 Algorithms

For a $K \times K$ sensor array, the filters $\{h_i\}_{i=0}^r$, $r = 2K - 1$, associated with the high-resolution image reconstruction are given by (3.2). With a proper assumption about boundary conditions, we have matrices $\{H_i\}_{i=0}^r$ corresponding to $\{h_i\}_{i=0}^r$, respectively. These matrices satisfy the perfect reconstruction formula as follows:

$$\sum_{i=0}^{r} H_i^T H_i = I. \tag{4.1}$$

We emphasize it again that (4.1) is the matrix representation of (2.4) with $\gamma = 0$ for the tight frame system arising from the high-resolution image reconstruction.

For the problem of high-resolution image reconstruction, the model (1.6) is

$$g = H_0 f + n,$$

where n is noise. With one step further, we consider a slightly more general restoration problem

$$g = H_s f + n \tag{4.2}$$

for a certain number s between 0 and r.

We propose two different types of algorithms for solving the problem (4.2). The algorithm of the first type is derived directly from the identity (4.1). It is basically the Landweber algorithm and will be presented in subsection 4.1. The algorithms of second type are modified versions of the algorithm of the first type by incorporating various denoising techniques. As we will see in subsection 4.2, some of algorithms are already appeared in our previous papers [9, 10, 11, 12, 13], but theoretical results on the convergence of those algorithms given here are new.

4.1   Basic algorithm

Multiplying a vector f from both sides of the perfect reconstruction formula (4.1) yields

$$f = H_s^T H_s f + \sum_{i \neq s} H_i^T H_i f. \tag{4.3}$$

By substituting the known data $H_s f \approx g$ in (4.2) into (4.3), we obtain an iteration as

$$f^{k+1} = H_s^T g + \sum_{i \neq s} H_i^T H_i f^k. \tag{4.4}$$

This is our basic algorithm. The proposed algorithms in the following subsection are all modification of (4.4) by incorporating different denoising schemes. Algorithm (4.4) is, in fact, no other than a Landweber's iteration. For the completeness, we give the convergence of (4.4) in the following theorem.

Theorem 1. *The sequence f^k generated by (4.4) with initial guess f^0 converges to a solution of*

$$\begin{cases} \min_f \|f - f^0\|_2, \\ s.t.\ H_s f = g. \end{cases} \tag{4.5}$$

Proof. First we write out the singular value decomposition of H_s, that is,

$$H_s = U_s \Sigma_s V_s^T,$$

where U_s and V_s are orthogonal matrices and Σ_s is a diagonal matrix. Without loss of generality, we assume that

$$\Sigma_s = \begin{bmatrix} \Sigma_{s,1} \\ & 0 \end{bmatrix}$$

with $\Sigma_{s,1}$ being an invertible diagonal matrix. The pseudo-inverse $H_s^\dagger$ of H_s reads

$$H_s^\dagger = V_s \Sigma_s^\dagger U_s^T,$$

where the pseudo-inverse $\Sigma_s^\dagger$ of Σ_s is given by

$$\Sigma_s^\dagger = \begin{bmatrix} \Sigma_{s,1}^{-1} \\ & 0 \end{bmatrix}.$$

Corresponding to the structure of Σ_s or $\Sigma_s^\dagger$, we can have a partition of any vector, say f, as follows

$$f = \begin{bmatrix} f_1 \\ f_2 \end{bmatrix},$$

where the dimension of f_1 is the same as the number of columns of $\Sigma_{s,1}$.

Now, we turn to the minimization problem (4.5). The constrained condition $H_s f = g$ implies

$$\Sigma_s V_s^T f = U_s^T g.$$

It, by the structure of Σ_s, says that

$$(V_s^T f)_1 = \Sigma_{s,1}^{-1}(U_s^T g)_1. \tag{4.6}$$

By using (4.6), we have

$$\begin{aligned}
\|f - f^0\|_2^2 &= \|V_s^T f - V_s^T f^0\|_2^2 \\
&= \|(V_s^T f)_1 - (V_s^T f^0)_1\|_2^2 + \|(V_s^T f)_2 - (V_s^T f^0)_2\|_2^2 \\
&= \|\Sigma_{s,1}^{-1}(U_s^T g)_1 - (V_s^T f^0)_1\|_2^2 + \|(V_s^T f)_2 - (V_s^T f^0)_2\|_2^2.
\end{aligned}$$

Obviously, to minimize $\|f - f^0\|_2^2$ with the constrained condition $H_s f = g$, the minimizer of (4.5), denoted by f^*, should satisfy the following conditions

$$(V_s^T f^*)_1 = \Sigma_{s,1}^{-1}(U_s^T g)_1 \quad \text{and} \quad (V_s^T f^*)_2 = (V_s^T f^0)_2. \tag{4.7}$$

Next, let us look at the iterative algorithm (4.4). Since $\sum_{i \neq s} H_s^T H_s = V_s(I - \Sigma_s^2)V_s^T$, then (4.4) becomes

$$f^{k+1} = V_s \Sigma_s U_s^T g + V_s(I - \Sigma_s^2)V_s^T f^k,$$

which is equivalent to

$$V_s^T f^{k+1} = \Sigma_s U_s^T g + (I - \Sigma_s^2)V_s^T f^k. \tag{4.8}$$

We can split (4.8) as

$$\begin{cases} (V_s^T f^{k+1})_1 = \Sigma_{s,1}(U_s^T g)_1 + (I - \Sigma_{s,1}^2)(V_s^T f^k)_1, \\ (V_s^T f^{k+1})_2 = (V_s^T f^k)_2. \end{cases} \tag{4.9}$$

Notice that the two iterations for $(V_s T f^k)_1$ and $(V_s T f^k)_2$ respectively are independent to each other. In the first equation of (4.9), the absolute values of all non-zero elements of the $\Sigma_{s,1}$ are strictly greater than 0 and less than or equal to 1. Therefore, the first iteration in (4.8) is a contract mapping, hence converges. Its limit $(V_s^T f^\star)_1$ is the unique fixed point of the first iteration in (4.9). More precisely, $(V_s^T f^\star)_1$ satisfies

$$(V_s^T f^\star)_1 = \Sigma_{s,1}(U_s^T g)_1 + (I - \Sigma_{s,1}^2)(V_s^T f^\star)_1,$$

which is equivalent to

$$(V_s^T f^\star)_1 = \Sigma_{s,1}^{-1}(U_s^T g)_1. \tag{4.10}$$

On the other hand, the second iteration in (4.9) converges obviously to

$$(V_s^T f^\star)_2 = (V_s^T f^0)_2. \tag{4.11}$$

Combining (4.10) and (4.11) together, we obtain (4.7). In other words, the limit of (4.4) is the solution of the minimization problem (4.5). □

Since (4.4) is a Landweber's iteration, it has a regularization property known as semiconvergence: the iterates f^k first approach the true image, but the noise will be amplified when k is larger than a certain threshold. Thus a stopping criterion, called the discrepancy principle, has to be introduced in order to obtain the best approximation of the required solution. Furthermore, the regularization property of projected Landweber's iterations is related to Tikhonov regularization, hence the edges in the underlying image are smeared. We further remark that the solution of (4.5) depends on the initial seed f^0.

4.2   Algorithms with tight frame denoising scheme

In this subsection, we will incorporate nonlinear denoising schemes into the iterative algorithm (4.4) with the aim of improving the quality of the reconstructed images. The Donoho's soft thresholding operator is adopted in our nonlinear denoising schemes. The soft-thresholding operator is given as follows

$$T_\lambda(u) = [t_{\lambda_1}(u_1), \cdots, t_{\lambda_n}(u_n)]^T \tag{4.12}$$

with λ being a pre-given vector having non-negative components and

$$t_{\lambda_j}(u_j) = \begin{cases} 0, & \text{if } |u_j| \le \lambda_j, \\ \operatorname{sgn}(u_j)(|u_j| - \lambda_j), & \text{if } |u_j| \le \lambda_j. \end{cases} \tag{4.13}$$

In what follows, the matrix A of the multilevel non-downsampled framelet decomposition is defined by (3.6) for the first two algorithm although it could be a matrix from any other tight frame system. We have following three modified algorithms.

4.2.1 Algorithm I

For framelet coefficients $H_i f^k$ of f^k in (4.4), we modify $H_i f^k$ into $A^T T_{\lambda_i}$ $(AH_i f^k)$ by using the multilevel non-downsampled framelet decomposition matrix A and the soft-thresholding operator T_{λ_i}. The resulting algorithm, called Algorithm I, from (4.4) is given as

$$f^{k+1} = H_s^T g + \sum_{i \neq s} H_i^T A^T T_{\lambda_i}(AH_i f^k). \qquad (4.14)$$

This algorithm was proposed and studied in [5, 8, 10, 11, 12, 13]. However, a complete analysis for this algorithm is not available. We will prove the convergence of this algorithm in the next section, and give minimization properties of its limit.

4.2.2 Algorithm II

We incorporate a frame based denoising scheme into each iterate of (4.4), and obtain Algorithm II as

$$f^{k+1} = A^T T_\lambda A\left(H_s^T g + \sum_{i \neq s} H_i^T H_i f^k\right). \qquad (4.15)$$

It was proposed and studied in [7, 8, 11], and can be seen as an extension of the algorithm in [16] from orthornormal system to tight frame systems; another possible extension can be found in [18]. The following convergence theorem for (4.15) is proved in [7, 8].

Theorem 2. *Let A be a tight frame. Let H_i, $0 \leq i \leq r$, be satisfying (4.3). Define*

$$\alpha^k = T_\lambda A\left(H_s^T g + \sum_{i \neq s} H_i^T H_i f^k\right), \qquad (4.16)$$

where f^k is generated by (4.15). Then α^k converges to a solution of

$$\min_\alpha \left\{\frac{1}{2}\|H_s A^T \alpha - g\|_2^2 + \frac{1}{2}\|(I - AA^T)\alpha\|_2^2 + \|\mathrm{diag}(\lambda)\alpha\|_1\right\}. \qquad (4.17)$$

Following [7, 8], the role of each terms in (4.17) can be explained as follows. The first term is the data fidelity, and the third term is to ensure the sparsity of the tight frame coefficient. The second term measures the distance from α to the range of A. By the framelet theory in [2, 24], if the coefficient α is in the range of A, then the (weighted) ℓ_1 norm α is equivalent to some Besov norm of the image $A^T \alpha$. Therefore, the second term links the (weighted) ℓ_1 norm to the regularity of the restored image in some sense. Combining all terms together, the minimization problem (4.17) fits the known data, and balances the sparsity of the frame coefficient and the regularity of restored image.

4.2.3   Algorithm III

With the same spirit of constructing matrix A using the piecewise linear tight frame in subsection 3.4, we design a matrix using the filters (3.2) associated with high-resolution image reconstruction. Such matrix, denoted by $\widetilde{A}$, can be written into

$$\widetilde{A} = [\widetilde{A}_{-1\to L}H_0, H_1, \cdots, H_s]^T, \tag{4.18}$$

where $\widetilde{A}_{-1\to L}$ is the framelet decomposition from level -1 to L. We consider the case that $s = 0$ and $s \neq 0$ respectively.

- If $s = 0$, then we have

$$\begin{aligned}
\widetilde{A}f &= [\widetilde{A}_{-1\to L}H_0 f, H_1 f, \cdots, H_r f]^T \\
&= [\widetilde{A}_{-1\to L}g, H_1 f, \cdots, H_r f]^T.
\end{aligned}$$

We obtain the algorithm

$$f^{k+1} = \widetilde{A}^T T_\lambda [\widetilde{A}_{-1\to L}g, H_1 f^k, \cdots, H_r f^k]^T. \tag{4.19}$$

- If $s \neq 0$, then we have

$$\begin{aligned}
\widetilde{A}f &= [\widetilde{A}_{-1\to L}H_0 f, H_1 f, \cdots, H_s f]^T \\
&= [\widetilde{A}_{-1\to L}H_0 f, H_1 f, \cdots, H_{s-1} f, g, H_{s+1} f, \cdots, H_r f]^T.
\end{aligned}$$

We obtain the algorithm

$$\begin{aligned}
f^{k+1} = \widetilde{A}^T T_\lambda [\widetilde{A}_{-1\to L}H_0 f^k, H_1 f^k, \cdots, \\
H_{s-1} f^k, g, H_{s+1} f^k, \cdots, H_r f^k]^T. \tag{4.20}
\end{aligned}$$

This algorithm can be found in [6, 7]. The following results are proved in [7] for the convergence of (4.19) and (4.20).

Theorem 3. *Let $\widetilde{A}$ be a tight frame in the form of* (4.18). *Define*

$$\alpha^k = T_\lambda [\widetilde{A}_{-1\to L}g, H_1 f^k, \cdots, H_r f^k]^T,$$

where f^k is generated by (4.19). *Then α^k converges to a solution of*

$$\min_{\alpha \in \mathcal{C}_0} \left\{ \|(I - \widetilde{A}\widetilde{A}^T)\alpha\|^2 + \|\mathrm{diag}(\lambda)\alpha\|_1 \right\}, \tag{4.21}$$

where $\mathcal{C}_0 = \{\alpha : \alpha|_\Gamma = T_{(\lambda|_\Gamma)}\widetilde{A}_{-1\to L}g\}$ with Γ being the positions corresponding to $\widetilde{A}_{-1\to J}H_0$ in $\widetilde{A}$.

Theorem 4. *Let $\widetilde{A}$ be a tight frame in the form of (4.18). Define*

$$\alpha^k = T_\lambda[\widetilde{A}_{-1 \to L} H_0 f^k, H_1 f^k, \cdots, H_{s-1} f^k, g, H_{s+1} f^k, \cdots, H_r f^k]^T,$$

where f^k is generated by (4.19). Then α^k converges to a solution of

$$\min_{\alpha \in \mathcal{C}_s} \{\|(I - \widetilde{A}\widetilde{A}^T)\alpha\|^2 + \|\mathrm{diag}(\lambda)\alpha\|_1\}, \qquad (4.22)$$

where $\mathcal{C}_s = \{\alpha : \alpha|_\Gamma = T_{(\lambda|_\Gamma)}g\}$ with Γ being the positions corresponding to H_s in $\widetilde{A}$.

Again, the minimization problems (4.21) and (4.22) fit the data, and balance the sparsity of the frame coefficient and the regularity of the restored image. The data fidelity comes from two aspects. On the one hand, it is obvious that the constraints $\alpha \in \mathcal{C}_0$ in (4.21) and $\alpha \in \mathcal{C}_s$ in (4.22) fit the data. On the other hand, the first term $\|(I - \widetilde{A}\widetilde{A}^T)\alpha\|^2$ in (4.21) and (4.22) also measures the data fidelity, since small $\|(I - \widetilde{A}\widetilde{A}^T)\alpha\|^2$ implies small $\|((I-\widetilde{A}\widetilde{A}^T)\alpha)|_\Gamma\|^2 \approx \|g-H_s\widetilde{A}^T\alpha\|^2$. The balance of the sparsity and the regularity comes from the first and second terms in (4.21) and (4.22) as we have explained for (4.17).

5 Analysis of Algorithm I

In this section, we focus on the convergence of algorithm (4.14). The main tool that we will use is an iterative algorithm of forward-backward splitting based on proximity operators.

5.1 Proximal forward-backward splitting

We first give a brief review of an iterative algorithm of forward-backward splitting based on proximity operators. This will be used in our analysis. The forward-backward splitting algorithm can be rooted back to 1950's for the study of numerical solution for partial differential equations. Here we are interested in a forward-backward splitting algorithm in [15] based on proximity operator. For any proper, lower semicontinuous, convex function $F(x)$ where $x \in \mathbb{R}^n$, the proximity operator prox_F [34, 35] is defined by

$$\forall y \in \mathbb{R}^n, \quad \mathrm{prox}_F(y) = \arg \min_x \left\{ \frac{1}{2}\|x - y\|_2^2 + F(x) \right\}, \qquad (5.1)$$

and Moreau's envelope, or Moreau-Yosida regularization [26], env_F is the continuously differentiable function

$$\forall y \in \mathbb{R}^n, \quad \mathrm{env}_F(y) = \min_x \left\{ \frac{1}{2}\|x - y\|_2^2 + F(x) \right\}. \qquad (5.2)$$

There is an important relation between the proximity operator and the envelope function, i.e.,

$$\nabla \operatorname{env}_F(y) = y - \operatorname{prox}_F(y). \tag{5.3}$$

Furthermore, both $\nabla \operatorname{env}_F$ and prox_F are all Lipshitz continuous with Lipshitz constant 1, i.e.,

$$\|(x - \operatorname{prox}_F(x)) - (y - \operatorname{prox}_F(y))\|_2 \le \|x - y\|_2, \quad \forall\, x, y \in \mathbb{R}^n,$$
$$\|\operatorname{prox}_F(x) - \operatorname{prox}_F(y)\|_2 \le \|x - y\|_2, \quad \forall\, x, y \in \mathbb{R}^n. \tag{5.4}$$

The proximal forward-backward splitting in [15] is to solve the minimization problem

$$\min_x \{F_1(x) + F_2(x)\}. \tag{5.5}$$

Here F_1 and F_2 are all proper, lower semicontinuous, convex functions. Moreover, the function F_2 is continuously differentiable, and its gradient is Lipschitz-continuous, i.e.,

$$\|\nabla F_2(x) - \nabla F_2(y)\|_2 \le \frac{1}{c}\|x - y\|_2. \tag{5.6}$$

With all the above settings, the iteration for solving (5.5) is

$$x^{k+1} = \operatorname{prox}_{dF_2}(x^k - d\nabla F_2(x^k)), \tag{5.7}$$

where $0 < d < 2c$. The convergence theory for (5.7) given in [15] is summarized as the following theorem.

Theorem 5. *Suppose that F_1 and F_2 satisfy*

1. *$F_1(x) + F_2(x)$ is coercive, i.e., whenever $\|x\|_2 \to +\infty$, $F_1(x) + F_2(x) \to +\infty$;*

2. *F_1 is a proper, convex, lower semi-continuous function; and*

3. *F_2 is a proper, convex, differentiable function satisfying (5.6) and $0 < d < 2c$.*

Then there exists at least one solution of (5.5), and for any initial guess x^0 the iteration (5.7) converges to a solution of (5.5).

5.2   Convergence of Algorithm I

Let

$$B = [AH_0; \ AH_1; \ \cdots; \ AH_r]. \tag{5.8}$$

By UEP and $A^T A = I$, we have $B^T B = I$. Therefore, the row vectors of B forms a tight frame. Since A is a framelet decomposition operator, B is a framelet package decomposition operator. Let

$$\alpha^k = [AH_0 f^k; \ \cdots; AH_{s-1}f^k; \ g; \ AH_{s+1}f^k; \ \cdots; \ AH_r f^k]$$

be the coefficients under the framelet package system B.

Define a new sequence of $(r+1)$-tuple as

$$\alpha^k = (T_\lambda AH_0 f^k, T_\lambda AH_1 f^k, \cdots, T_\lambda AH_{s-1} f^k, g,$$
$$T_\lambda AH_{s+1} f^k, \cdots, T_\lambda AH_r f^n), \tag{5.9}$$

and its i-th entry, $i = 0, 1, \cdots, r$, are subscripted by i, i.e., $\alpha_s^k = g$ and $\alpha_i^k = AH_i f^k$ for $i \neq s$ and $i = 1, \cdots, r$. We write a new matrix B as

$$B = [AH_0; AH_1; \cdots; AH_{s-1}; H_s; AH_{s+1}; \cdots; AH_r]^T.$$

Note that B and B^T correspond to a framelet packet decomposition and reconstruction operator respectively, see [8]. Since $A^T A = I$ and $\sum_{i=0}^r H_i^T H_i = I$, we have $B^T B = I$. By (4.14) and (5.9), we obtain

$$f^{k+1} = B^T \alpha^k. \tag{5.10}$$

Lemma 1. *The sequence α^k converges if and only if the sequence f^k does.*

Proof. Since all the operators involved in (5.9) and (5.10) are all continuous, the lemma follows immediately. $\square$

Therefore, we show that f^k converges by showing that α^k converges. The strategy is the following: we first transform the iteration into a proximal forward-backward splitting iteration for a special functional; then by applying Theorem 5, we obtain the convergence for the sequence α^k.

Let $\mathcal{C} = \{\alpha | \alpha_s = g\}$ which is a subset of the sub of all $(r+1)$-tuples. It is obviously a closed nonempty convex set. We define the indicator function of $\mathcal{C}$ by

$$D_\mathcal{C}(\alpha) = \begin{cases} 0, & \alpha \in \mathcal{C}, \\ \infty, & \alpha \notin \mathcal{C}. \end{cases}$$

Lemma 2. *The sequence α^k defined in (5.9) generated by algorithm (4.14) is equivalent to that generated by a proximal forward-backward splitting iteration (5.7) with $d = 1$ for the minimization problem*

$$\min_{\alpha \in \mathcal{C}} \left\{ \|(I - BB^T)\alpha\|_2^2 + \sum_{i \neq s} \|\mathrm{diag}(\lambda)\alpha_i\|_1 \right\}, \tag{5.11}$$

where

$$F_1(\alpha) = D_\mathcal{C}(\alpha) + \sum_{i \neq s} \|\mathrm{diag}(\lambda)\alpha_i\|_1 \text{ and } F_2(\alpha) = \|(I - BB^T)\alpha\|_2^2.$$

Proof. By (4.14) and (5.9), we have

$$\alpha_s^{k+1} = g \quad \text{and} \quad \alpha_i^{k+1} = T_\lambda [BB^T \alpha^k]_i, \quad i \neq s. \tag{5.12}$$

Here, $[\cdot]_i$ denotes the i-th entry of the $(r+1)$-tuple. It is clear that

$$\nabla F_2(\alpha) = \alpha - BB^T \alpha. \tag{5.13}$$

Comparing (5.12) and (5.13) with (5.7) where $d = 1$, we see that if we can prove that

$$\mathrm{prox}_{F_1}(\alpha) = (T_\lambda \alpha_0, \cdots, T_\lambda \alpha_{s-1}, g, T_\lambda \alpha_{s+1}, \cdots, T_\lambda \alpha_r), \tag{5.14}$$

then we are done. We verify (5.14) by considering the definition of the proximal operator in the following

$$\mathrm{prox}_{F_1}(\alpha) = \arg \min_{\beta} \left\{ \frac{1}{2} \|\beta - \alpha\|_2^2 + \frac{1}{2} D_\mathcal{C}(\beta) + \frac{1}{2} \sum_{i \neq s} \|\mathrm{diag}(\lambda)\beta_i\|_1 \right\}. \tag{5.15}$$

Note that $D_\mathcal{C}(\alpha) = \sum_{i=1}^n D_{g_i}(\alpha_{i,j})$, where $D_{g_i}(\alpha_{i,s}) = 0$ if $g_i = \alpha_{i,s}$ and $D_{g_i}(\alpha_{i,j}) = \infty$ otherwise. Therefore, by (5.15),

$$[\mathrm{prox}_{F_1}(\alpha)]_{i,s} = \arg \min_{\beta_{i,s}} \left\{ \frac{1}{2}(\beta_{i,s} - \alpha_{i,s})^2 + \frac{1}{2} D_{g_i}(\beta_{i,s}) \right\}, \quad i = 1, \cdots, n,$$

and

$$[\mathrm{prox}_{F_1}(\alpha)]_{i,j} = \arg \min_{\beta_{i,j}} \left\{ \frac{1}{2}(\beta_{i,j} - \alpha_{i,j})^2 + \frac{1}{2} \lambda_i |\beta_{i,j}| \right\}, \quad j \neq s.$$

The above two equations are identical to (5.14). $\square$

By applying Theorem 5, we get that the sequence α^k converges to a minimizer of (5.11).

Theorem 6. *The sequence α^k generated by (5.9) converges to a solution of (5.11).*

Proof. We note that the set $\mathcal{C}$ is a closed non-empty convex set, hence $D_\mathcal{C}$ is a proper lower semi-continuous convex function. Therefore, both the functions F_1 and F_2 are proper, semi-continuous and convex, and F_2 is differentiable. The gradient of F_2 is Lipschitz continuous with Lipschitz constant 1. Indeed, for any $(r+1)$-tuple α and β, we have

$$\|\nabla F_2(\alpha) - \nabla F_2(\beta)\|_2 = \|(I - BB^T)(\alpha - \beta)\|_2 \leq \|I - BB^T\|_2 \|\alpha - \beta\|_2.$$

Since the operator $(I - BB^T)$ satisfies $(I - BB^T)^2 = (I - BB^T)$ and $(I - BB^T)^T = (I - BB^T)$, it is an orthonormal projector, hence its norm is 1. Therefore, ∇F_2 is Lipschitz continuous with Lipschitz constant 1, hence $c = 1$. Therefore, $d = 1$ satisfies the condition in Theorem 5. Thus, by Theorem 5, if we can prove the existence of the minimizers for (5.11), we get the lemma.

We show the existence by showing the coercivity of F_1 hence of $F_1 + F_2$. If $\alpha \in \mathcal{C}$, then we have

$$F_1(\alpha) = \sum_{i \neq s} \|\mathrm{diag}(\lambda)\alpha_i\|_1 \geq \min_j \lambda_j \sqrt{(\sum_{i \neq s} \|\alpha_i\|_2^2)} = \min_j \lambda_j (\|\alpha\|_2^2 - \|g\|_2^2).$$

If $\alpha \notin \mathcal{C}$, then it is clear $F_1(\alpha) = \infty$. Therefore, as $\|\alpha\|_2 \to \infty$, $F_1(\alpha) \to \infty$. Hence, F_1 is coercive. □

5.3 Minimization property of Algorithm I

We have proven that the limit α^k generated by (5.9) is a solution of (5.11). Let f^* and α^* be the limit of the sequence f^k and α^k respectively. Then by (5.9), we have that $f^* = B^T \alpha^*$. The following lemma explains the minimization problem (5.11). It states the optimality property of the pair $\{f^*, \alpha^*\}$ in the sense that, if f^* is perturbed by e and α^* is perturbed by Be, then the energy $\|H_s f^* - g\|_2^2 + \sum_{i \neq s} \|\mathrm{diag}(\lambda)\alpha_s^*\|_1$ increases. The first term in the energy is a data fidelity, and the second term is a weighted ℓ_1 norm which leads to the sparsity.

Lemma 3. *For any vector* $e \in \mathbb{R}^n$, *we have*

$$\|H_s(f^* + e) - g\|_2^2 + \sum_{i \neq s} \|\mathrm{diag}(\lambda)(AH_i e + \alpha_s^*)\|_1$$

$$\geq \|H_s f^* - g\|_2^2 + \sum_{i \neq s} \|\mathrm{diag}(\lambda)\alpha_s^*\|_1. \tag{5.16}$$

Proof. Let

$$\gamma := (AH_0 e, \cdots, AH_{s-1} e, 0, AH_{s+1} e, \cdots, AH_r e).$$

Since α^* is a minimizer of (5.11), we have

$$\|(I - BB^T)(\gamma + \alpha^*)\|_2^2 + \sum_{i \neq s} \|\mathrm{diag}(\lambda)(AH_i e + \alpha_i^*)\|_1$$

$$\geq \|(I - BB^T)\alpha^*\|_2^2 + \sum_{i \neq s} \|\mathrm{diag}(\lambda)\alpha_i^*\|_1. \tag{5.17}$$

By the definition of B, the $(r + 1)$-tuple

$$\delta = (AH_0 e, \cdots, AH_{s-1} e, H_s e, AH_{s+1} e, \cdots, AH_r e)$$

satisfies $\delta = Be$. Therefore, by the identity $B^T Be = e$ we obtain

$$(I - BB^T)\delta = 0, \tag{5.18}$$

which implies that

$$\|(I - BB^T)(\gamma + \alpha^\star)\|_2^2 = \|(I - BB^T)(\gamma - \delta + \alpha^\star)\|_2^2. \tag{5.19}$$

Notice that

$$B^T(\gamma - \delta) = H_s^T H_s e. \tag{5.20}$$

Substituting (5.20) and the definition of $f^\star = A^T \alpha^\star$ into (5.19), we have that

$$\|(I - BB^T)(\gamma + \alpha^\star)\|_2^2 = \|(\alpha^\star - Bf^\star) + (\gamma - \delta) - BH_s^T H_s e\|_2^2, \tag{5.21}$$

which is equivalent to

$$\begin{aligned}
&\|(I - BB^T)(\gamma + \alpha^\star)\|_2^2 \\
&= \|(\alpha^\star - Bf^\star) + (\gamma - \delta)\|_2^2 + \|BH_s^T H_s e\|_2^2 \\
&\quad -2\langle(\alpha^\star - Bf^\star) + (\gamma - \delta), BH_s^T H_s e\rangle \\
&= \|(\alpha^\star - Bf^\star) + (\gamma - \delta)\|_2^2 + \|BH_s^T H_s e\|_2^2 \\
&\quad -2\langle B^T(\alpha^\star - Bf^\star) + B^T(\gamma - \delta), H_s^T H_s e\rangle.
\end{aligned} \tag{5.22}$$

Since $B^T B = I$, we have

$$\|BH_s^T H_s e\|_2^2 = \|H_s^T H_s e\|_2^2 \tag{5.23}$$

and

$$B^T(\alpha^\star - Bf^\star) = B^T\alpha^\star - f^\star = 0. \tag{5.24}$$

Substituting (5.23) and (5.24) into (5.22), we get

$$\begin{aligned}
\|(I - BB^T)(\gamma + \alpha^\star)\|_2^2 = \|(\alpha^\star - Bf^\star) + (\gamma - \delta)\|_2^2 + \|H_s^T H_s e\|_2^2 \\
-2\langle B^T(\gamma - \delta), H_s^T H_s e\rangle,
\end{aligned}$$

which together with (5.20) implies

$$\begin{aligned}
&\|(I - BB^T)(\gamma + \alpha^\star)\|_2^2 \\
&= \|(\alpha^\star - Bf^\star) + (\gamma - \delta)\|_2^2 + \|H_s^T H_s e\|_2^2 - 2\langle H_s^T H_s e, H_s^T H_s e\rangle \\
&= \|(\alpha^\star - Bf^\star) + (\gamma - \delta)\|_2^2 - \|H_s^T H_s e\|_2^2.
\end{aligned} \tag{5.25}$$

Since the first term in the last equality can be rewritten into

$$\|(\alpha^\star - Bf^\star) + (\gamma - \delta)\|_2^2 = \sum_{i \neq s} \|\alpha_i^\star - AH_i f^\star\|_2^2 + \|g - H_s(f^\star + e)\|_2^2,$$

we obtain

$$\|(I - BB^T)(\gamma + \alpha^\star)\|_2^2$$
$$= \sum_{i \neq s} \|\alpha_i^\star - AH_i f^\star\|_2^2 + \|g - H_s(f^\star + e)\|_2^2 - \|H_s^T H_s e\|_2^2. \quad (5.26)$$

On the other hand, we have

$$\|(I - BB^T)\alpha^\star\|_2^2 = \sum_{i \neq s} \|\alpha_i^\star - AH_i f^\star\|_2^2 + \|g - H_s f^\star\|_2^2. \quad (5.27)$$

Substituting (5.26) and (5.27) into (5.18), we obtain

$$\|g - H_s(f^\star + e)\|_2^2 - \|H_s^T H_s e\|_2^2 + \sum_{i \neq s} \|\mathrm{diag}(\lambda)(AH_i e + \alpha_i^\star)\|_1$$

$$\geq \|g - H_s f^\star\|_2^2 + \sum_{i \neq s} \|\mathrm{diag}(\lambda)\alpha^\star\|_1,$$

which obviously implies (5.11). □

6 Numerical experiments

We now provide numerical experiments of the presented Algorithm I, Algorithm II, and Algorithm III for high-resolution image reconstruction. The algorithms are evaluated by using peak signal-to-noise ratio (PSNR) which compares the reconstructed image f_c with the original image f. It is defined by

$$10 \log_{10} \frac{255^2 M_1 M_2}{\|f - f_c\|_2^2}$$

where the size of the reconstructed images is $M_1 \times M_2$.

We use the "Cameraman" and "Bridge" images of size 256×256 as the original images, see Figure 6.1. The maximum number of iteration

Figure 6.1 Original "Bridge" image (left) and original "Cameranman" images (right).

is set to 100 and the iteration processes of Algorithm I, Algorithm II, and Algorithm III are stopped when the reconstructed images achieve the highest PSNR values.

For a 2×2 sensor array, the observed images are shown in the top row of Figure 6.2. The reconstructed "Bridge" images (from the second row to the fourth row, the left column of Figure 6.2) via Algorithm I, Algorithm II, and Algorithm III have PSNR values of 26.10 dB, 26.63 dB, and 25.97 dB, respectively. Likewise, the reconstructed "Cameraman" images (from the second row to the fourth row, the right column of Figure 6.2) via Algorithm I, Algorithm II, and Algorithm III have PSNR values of 31.75 dB, 31.16 dB, and 31.68 dB, respectively.

Similar experiments are conducted for "Bridge" and "Cameraman" images in 4×4 sensor array. The reconstructed "Bridge" images (from the second row to the fourth row, the left column of Figure 6.3) via Algorithm I, Algorithm II, and Algorithm III have PSNR values of 23.71dB, 24.00dB, and 23.45 dB, respectively. Likewise, the reconstructed "Cameraman" images (from the second row to the fourth row, the right column of Figure 6.2) via Algorithm I, Algorithm II, and Algorithm III have PSNR values of 27.56 dB, 27.38 dB, and 27.56 dB, respectively.

In summary, the proposed algorithms have produced very similar results for high-resolution image reconstruction in terms of the PSNR values and the visual quality of the reconstructed images. However, from the viewpoint of computational efficiency, we recommend to choose Algorithm II for the problem of high-resolution image reconstruction.

Figure 6.2   2 × 2 sensor array. Left column (from top to bottom) shows the observed "Bridge" image, reconstructed images by Algorithm I, Algorithm II, and Algorithm III with PSNR values of 26.10dB, 26.63dB, and 25.97 dB, respectively. Likewise, right column (from top to bottom) shows the observed "Cameraman" image, reconstructed images by Algorithm I, Algorithm II, and Algorithm III with PSNR values of 31.75dB, 31.16dB, and 31.68 dB.

Figure 6.3   4 × 4 sensor array. Left column (from top to bottom) shows the observed "Bridge" image, reconstructed images by Algorithm I, Algorithm II, and Algorithm III with PSNR values of 23.71dB, 24.00dB, and 23.45 dB, respectively. Likewise, right column (from top to bottom) shows the observed "Cameraman" image, reconstructed images by Algorithm I, Algorithm II, and Algorithm III with PSNR values of 27.56dB, 27.38dB, and 27.56 dB.

References

[1] S. BORMAN AND R. STEVENSON, *Super-resolution from image sequences — a review*, in Proceedings of the 1998 Midwest Symposium on Circuits and Syatems, vol. 5 (1998).

[2] L. BORUP, R. GRIVONBAL, AND M. NIELSEN, *Bi-framelet systems with few vanishing moments characterize Besov spaces*, Applied and Computational Harmonic Analysis, 17 (2004), pp. 3–28.

[3] N. BOSE AND K. BOO, *High-resolution image reconstruction with multisensors*, International Journal of Imaging Systems and Technology, 9 (1998), pp. 294–304.

[4] N. BOSE, S. LERTRATTANAPANICH, AND J. KOO, *Advances in superresolution using the l-curve*, vol. II of Proc. Int. Symp. Circuits and Systems, Sydney, NSW, Australia (2001).

[5] J.-F. CAI, R. CHAN, L. SHEN, AND Z. SHEN, *Restoration of chopped and nodded images by framelets*, SIAM Journal on Scientific Computing, 30 (2008), no. 3, pp. 1205–1227.

[6] J.-F. CAI, R. H. CHAN, L. SHEN, AND Z. SHEN, *Simultaneously inpainting in image and transformed domains*, Numerische Mathematik, 112 (2009), no. 4, pp. 509–533.

[7] J.-F. CAI AND Z. SHEN, *Deconvolution: A wavelet frame approach, II*, preprint (2008).

[8] A. CHAI AND Z. SHEN, *Deconvolution: A wavelet frame approach*, Numerische Mathematik, 106 (2007), pp. 529–587.

[9] R. CHAN, T. CHAN, L. SHEN, AND Z. SHEN, *Wavelet algorithms for high-resolution image reconstruction*, SIAM Journal on Scientific Computing, 24 (2003), pp. 1408–1432.

[10] ——, *Wavelet deblurring algorithms for spatially varying blur from high-resolution image reconstruction*, Linear Algebra and its Applications, 366 (2003), pp. 139–155.

[11] R. CHAN, S. D. RIEMENSCHNEIDER, L. SHEN, AND Z. SHEN, *High-resolution image reconstruction with displacement errors: A framelet approach*, International Journal of Imaging Systems and Technology, 14 (2004), pp. 91–104.

[12] ——, *Tight frame: The efficient way for high-resolution image reconstruction*, Applied and Computational Harmonic Analysis, 17 (2004), pp. 91–115.

[13] R. CHAN, Z. SHEN, AND T. XIA, *A framelet algorithm for enchancing video stills*, Applied and Computational Harmonic Analysis, 23 (2007), pp. 153–170.

[14] C. CHUI, W. HE, AND J. STOCKLER, *Compactly supported tight and sibling frames with maximum vanishing moments*, Applied and Computation Harmonic Analysis, 13 (2002), pp. 224–262.

[15] P. COMBETTES AND V. WAJS, *Signal recovery by proximal forward-backward splitting*, Multiscale Modeling and Simulation: A SIAM Interdisciplinary Journal, 4 (2005), pp. 1168–1200.

[16] I. DAUBECHIES, M. DEFRISE, AND C. D. MOL, *An iterative thresholding algorithm for linear inverse problems with a sparsity constraint*, Communications on Pure and Applied Mathematics, 57 (2004), pp. 1413–1541.

[17] I. DAUBECHIES, B. HAN, A. RON, AND Z. SHEN, *Framelets: MRA-based constructions of wavelet frames*, Applied and Computation Harmonic Analysis, 14 (2003), pp. 1–46.

[18] I. DAUBECHIES, G. TESCHKE, AND L. VESE, *Iteratively solving linear inverse problems under general convex constraints*, Inverse Problems and Imaging, 1 (2007), pp. 29–46.

[19] C. DE BOOR, R. DE VORE, AND A. RON, *On the construction of multivariate (pre)-wavelets*, Constructive Approximation, 9 (1993), pp. 123–166.

[20] R. DUFFIN AND A. SCHAEFFER, *A class of nonharmonic Fourier series*, Transactions on American Mathematics Society, 72 (1952), pp. 341–366.

[21] M. ELAD AND A. FEUER, *Restoration of a single superresolution image from several blurred, noisy and undersampled measured images*, IEEE Transactions on Image Processing, 6 (1997), pp. 1646–1658.

[22] ———, *Superresolution restoration of an image sequence: adaptive filtering approach*, IEEE Transactions on Image Processing, 8 (1999), pp. 387–395.

[23] M. ELAD AND Y. HEL-OR, *A fast super-resolution reconstruction algorithm for pure translational motion and common space-invariant blur*, IEEE Transactions on Image Processing, 10 (2001), pp. 1187–1193.

[24] B. HAN AND Z. SHEN, *Dual wavelet frames and Riesz bases in Sobolev spaces*, Constructive Approximation, 29 (2009), no. 3, pp. 369–406.

[25] R. HARDIE, K. BARNARD, AND E. ARMSTRONG, *Joint MAP registration and high-resolution image estimation using a sequence of undersampled images*, IEEE Transactions on Image Processing, 6 (1997), pp. 1621–1633.

[26] J.-B. HIRIART-URRUTY AND C. LEMARECHAL, *Convex analysis and minimization algorithms*, vol. 305 of Grundlehren der Mathematischen Wissenschaften [Fundamental Principles of Mathematical Sciences], Springer-Verlag, Berlin, 1993.

[27] M. HONG, M. KANG, AND A. KATSAGGELOS, *An iterative weighted regularized algorithm for improving the resolution of video sequences*, in IEEE International Conference On Image Processing, 1997.

[28] T. HUANG AND R. TSAY, *Multiple frame image restoration and registration*, in Advances in Computer Vision and Image Processing, T. S. Huang, ed., vol. 1, Greenwich, CT: JAI (1984), pp. 317–339.

[29] R. JIA AND Z. SHEN, *Multiresolution and wavelets*, Proceedings of the Edinburgh Mathematical Society, 37 (1994), pp. 271–300.

[30] S. P. KIM, N. K. BOSE, AND H. M. VAKENZUELA, *Recursive reconstruction of high resolution image from noisy undersampled multiframes*, IEEE Transactions on Acoustics, Speech, and Signal Processing, 38 (1990), pp. 1013–1027.

[31] T. KOMATSU, K. AIZAWA, T. IGARASHI, AND T. SAITO, *Signal-processing based method for acquiring very high resolution image with multiple cameras and its theoretical analysis*, IEE Proceedings: Communications, Speech and Vision, 140 (1993), pp. 19–25.

[32] Y. LU, L. SHEN, AND Y. XU, *Multi-parameter regularization methods for high-resolution image reconstruction with displacement errors*, IEEE Transactions on Circuit and System I: Fundamental Theory and Applications, 54 (2007), pp. 1788–1799.

[33] R. MOLINA, M. VEGA, J. ABAD, AND A. KATSAGGELOS, *Parameter estimation in Bayesian high-resolution image reconstruction with multisensors*, IEEE Transactions on Image Processing, 12 (2003), pp. 1655–1667.

[34] J.-J. MOREAU, *Fonctions convexes duales et points proximaux dans un espace hilbertien*, C.R. Acad. Sci. Paris Sér. A Math., 255 (1962), pp. 1897–2899.

[35] ———, *Proximité et dualité dans un espace hilbertien*, Bull. Soc. Math. France, 93 (1965), pp. 273–299.

[36] M. NG, R. CHAN, T. CHAN, AND A. YIP, *Cosine transform preconditioners for high resolution image reconstruction*, Linear Algebra and its Applications, 316 (2000), pp. 89–104.

[37] M. NG AND A. YIP, *A fast MAP algorithm for high-resolution image reconstruction with multisensors*, Multidimensional Systems and Signal Processing, 12 (2001), pp. 143–164.

[38] M. K. NG AND N. BOSE, *Analysis of displacement errors in high-resolution image reconstruction with multisensors*, IEEE Transactions on Circuits and Systems — I: Fundamental Theory and Applications, 49 (2002), pp. 806–813.

[39] M. K. NG, J. KOO, AND N. BOSE, *Constrained total least squares computations for high resolution image reconstruction with multisensors*, International Journal of Imaging Systems and Technology, 12 (2002), pp. 35–42.

[40] N. NGUYEN AND P. MILANFAR, *A wavelet-based interpolation-restoration method for superresolutio*, IEEE Transactions on Circuits, Systems, and Signal Processing, 19 (2000), pp. 321–338.

[41] N. NGUYEN, P. MILANFAR, AND G. GOLUB, *A computationally efficient superresolution image reconstruction algorithm*, IEEE Transactions on Image Processing, 10 (2001), pp. 573–583.

[42] A. RON AND Z. SHEN, *Affine system in $L_2(R^d)$: the analysis of the analysis operator*, Journal of Functional Analysis, 148 (1997), pp. 408–447.

[43] R. SCHULTZ AND R. STEVENSON, *Extraction of high-resolution frames from video sequences*, IEEE Transactions on Image Processing, 5 (1996), pp. 996–1011.

[44] L. SHEN AND Q. SUN, *Bi-orthogonal wavelet system for high-resolution image reconstruction*, IEEE Transactions on Signal Processing, 52 (2004), pp. 1997–2011.

[45] B. TOM, N. GALATSANOS, AND A. KATSAGGELOS, *Reconstruction of a high-resolution image from multiple low resolution images*, Super-Resolution Imaging, Kluwer Academic Publisher, 2001, ch. 4, pp. 73–105.

Greedy Algorithms for Adaptive Triangulations and Approximations

Albert Cohen

Laboratoire Jacques-Louis Lions
Université Pierre et Marie Curie, France
Email: cohen@ann.jussieu.fr

Abstract

We consider two problems of non-linear approximation in which an arbitrary function f is respectively approximated by an N-term combination from a dictionary and a piecewise affine function on N triangles. For both problems we present simple greedy algorithms which aim to provide with near-optimal solutions at a reasonable computational cost. We discuss the rate of convergence for these algorithms.

1   Introduction

In approximation theory, one studies the process of approaching arbitrary functions by simple functions depending on N parameters, such as algebraic or trigonometric polynomials, finite elements or wavelets. One usually makes the distinction between linear and nonlinear approximation. In the first case, the simple function is picked from a linear space (such as polynomials of degree N or piecewise constant functions on some fixed partition of cardinality N) and is typically computed by projection of the arbitrary function onto this space. In the second case, the simple function is picked from a nonlinear space, yet still characterizable by N parameters. Such a situation typically occurs when dealing with adaptive or data driven approximations, which makes it relevant for applications as diverse as data compression, statistical estimation or numerical schemes for partial differential or integral equations (see [16] for a general survey). However the notion of projection is not anymore applicable and therefore a critical question is: *how to compute the best possible approximation to a given function?* Let us translate this question in concrete terms for two specific examples.

Best N-term approximation: given a dictionary $\mathcal{D}$ of functions which is normalized and complete in some Banach space $\mathcal{H}$, and given $f \in \mathcal{H}$

and $N > 0$, find the combination $f_N = \sum_{k=1}^N c_k g_k$ which approximates f at best, where $\{c_1, \cdots, c_N\}$ are real numbers and $\{g_1, \cdots, g_N\}$ are picked from $\mathcal{D}$.

Adaptive triangulations: given a function f defined on a polygonal domain and given $N > 0$, find a partition of the domain into N triangles such that the error in a prescribed norm between f and its projection onto piecewise polynomial functions of some fixed degree on this partition is minimized.

In order to make these problems computationally tractable, one may assume in the first example that the search is limited to a subset of $\mathcal{D}$ of cardinality M, or in the second example that the vertices of each triangle are picked within a limited yet large number M of locations. However the exhaustive search for the optimal solution has the combinatorial order of complexity $\binom{M}{N}$ and both problems are therefore generally not solvable in polynomial time in N and M. A relevant goal is therefore to look for sub-optimal yet acceptable solutions which can be computed in reasonable time.

Greedy algorithms constitute a simple approach for achieving this goal. They rely on stepwise local optimization procedures for picking the parameters of the approximant in an inductive fashion, with the hope of approaching the globally optimal solution. They are particularly easy to implement, yet the analysis of their approximation performance gives rise to many open problems.

This paper is to present this type of algorithms in the framework of the two above mentioned examples and discuss their properties. Best N-term approximations and adaptive triangulations are respectively discussed in sections 2 and 3.

2   Best N-term approximation

We first address the best N-term approximation problem. For the sake of simplicity, we shall assume here that we work in a Hilbert space $\mathcal{H}$, although Banach space have also been considered, see in particular [23, 24]. We shall assume that the dictionary $\mathcal{D}$ is complete in $\mathcal{H}$ and that all elements are normalized in the sense that

$$\|g\|_{\mathcal{H}} = 1.$$

Before describing several greedy algorithms which have been proposed in this context, it is instructive to consider the simpler case where the dictionary $\mathcal{D}$ is an orthonormal basis of $\mathcal{H}$.

2.1   The case of an orthonormal basis

When $\mathcal{D}$ is an orthonormal basis, the exact best N-term approximation can be computed by an elementary algorithm which retains the the N largest coefficients $c_g = \langle f, g \rangle$ in the expansion of f, i.e. defining

$$f_N = \sum_{N \text{ largest } |c_g|} c_g g.$$

Intuitively, this approximation process is effective when the coefficient sequence $(c_g)_{g \in \mathcal{D}}$ is *concentrated* or *sparse*. One way to measure sparsity is by introducing the smallest value of $0 < p \leq 2$ such that

$$\#\{g : |c_g| \geq \eta\} \leq C\eta^{-p},$$

for some fixed C independent of $\eta > 0$, i.e. such that $(c_g)_{g \in \mathcal{D}}$ is weakly ℓ^p-summable (or belongs to $w\ell^p(\mathcal{D})$), the extreme case $p = 0$ corresponding to a finitely supported sequence. This property can also be expressed by saying that if we reorder the coefficients in decreasing order of magnitude, the resulting sequence $(c_n^*)_{n>0}$ decays like $n^{-\frac{1}{p}}$. Note that

$$\|f - f_N\|_{\mathcal{H}} = \Big(\sum_{n>N} (c_n^*)^2 \Big)^{\frac{1}{2}},$$

and therefore if $(c_g)_{g \in \mathcal{D}} \in w\ell^p$ we obtain that

$$\|f - f_N\|_{\mathcal{H}} \leq CN^{-s}, \quad s = \frac{1}{p} - \frac{1}{2}.$$

It can easily be checked that this rate of decay is equivalent to the property $(c_g)_{g \in \mathcal{D}} \in w\ell^p$.

In summary the convergence rate s of best N-term approximation is directly governed by the degree of sparsity p. A natural question is therefore: What properties of f guarantee a certain degree of sparsity in its expansion. The answer is of course depending on the basis under consideration. For instance, sparsity in a wavelet basis is equivalent to certain smoothness properties measured in the scale of Besov space. Other type smoothness properties govern sparsity in Fourier series or Legendre polynomial basis.

In the following we want to remain in the abstract setting where we do not specify the basis, and turn to the situation where this basis is now replaced by a non-orthogonal and possibly redundant dictionary. There are many applications where this situation is relevant, in particular in signal and image processing and in statistical learning.

2.2   Greedy algorithm for N-term approximation

Greedy algorithms for building N-term approximations were initially introduced in the context of statistical data analysis. Their approximation properties were first explored in [5, 19] in relation with neural network estimation, and in [17] for general dictionaries. Surveys on such algorithms is given in [23, 24].

We only describe here the four most commonly used greedy algorithms:

1. Stepwise Projection (SP): $\{g_1, \cdots, g_{k-1}\}$ being selected we define f_{k-1} as the orthogonal projection onto $\mathrm{Span}(g_1, \cdots, g_{k-1}\}$. The next g_k is selected so to minimize the distance between f and $\mathrm{Span}(g_1, \cdots, g_{k-1}, g\}$ among all choices of $g \in \mathcal{D}$.

2. Orthonormal Matching Pursuit (OMP): with the same definition for f_{k-1}, we select g_k so to maximize the inner product $|\langle f - f_{k-1}, g\rangle|$ among all choices of $g \in \mathcal{D}$. In contrast to SP, we do not need to evaluate the anticipated projection error for all choices of $g \in \mathcal{D}$, which makes OMP more attractive from a computational viewpoint.

3. Relaxed Greedy Algorithm (RGA): f_{k-1} being constructed, we define $f_k = \alpha_k f_{k-1} + \beta_k g_k$, where (α_k, β_k, g) are selected so to minimize the distance between f and $\alpha f_{k-1} + \beta g$ among all choices of (α, β, g). It is often convenient to fix α_k in advance, which leads to selecting g_k which maximizes $|\langle f - \alpha_k f_{k-1}, g\rangle|$ and $\beta_k = \langle f - \alpha_k f_{k-1}, g_k\rangle$. A frequently used choice is $\alpha_k := (1 - c/k)_+$ for some fixed $c > 1$. The intuitive role of the relaxation parameter α_k is to damp the memory of the algorithm which might have been misled in its first steps. Since no orthogonal projection is involved, RGA is even cheaper than OMP.

4. Pure Greedy Algorithm (PGA): this is simply RGA with the particular choice $\alpha_k = 1$. We therefore select g_k so to maximize the inner product $|\langle f - f_{k-1}, g\rangle|$ as in OMP, and then set $f_k = f_{k-1} + \langle f - f_{k-1}, g_k\rangle g_k$.

It should be noted that in the case where $\mathcal{D}$ is an orthonormal basis, SP, OMP and PGA are equivalent to the procedure of retaining the largest coefficients in the expansion of f which is known to produce the best N-term approximation.

2.3   Approximation results

For a general dictionary, $\mathcal{D}$ a natural question is wether a similar property holds: if f admits a sparse representation in $\mathcal{D}$, can we derive some corresponding rate of convergence for the greedy algorithm? By analogy with the case of an orthonormal basis, we could assume that $f = \sum_{g \in \mathcal{D}} c_g g$ with $(c_g)_{g \in \mathcal{D}} \in w\ell^p$ and ask wether the greedy algorithm converges with rate N^{-s} with $s = \frac{1}{p} - \frac{1}{2}$. However, the condition $(c_g)_{g \in \mathcal{D}} \in w\ell^p$ is not anymore appropriate since it does not generally guarantee the convergence of $\sum_{g \in \mathcal{D}} c_g g$ in $\mathcal{H}$.

A first set of results concerns the case where f admits a summable expansion, i.e. $(c_g)_{g \in \mathcal{D}} \in \ell^1$ or equivalently f belongs to a multiple of the convex hull of $(-\mathcal{D}) \cup \mathcal{D}$. In this case, the series $\sum_{g \in \mathcal{D}} c_g g$ converges in $\mathcal{H}$ by triangle inequality. We denote as $\mathcal{L}^1$ the space of such f, equipped with the norm

$$\|f\|_{\mathcal{L}^1} := \inf_{f = \sum c_g g} \sum_{g \in \mathcal{D}} |c_g|.$$

Clearly $\mathcal{L}^1 \subset \mathcal{H}$ with continuous embedding. The following result was proved in [19] for SP and RGA with the choice $\alpha_k := (1 - c/k)_+$ and in [17] for OMP.

Theorem 2.1. *If $f \in \mathcal{L}^1$, then*

$$\|f - f_N\|_{\mathcal{H}} \leq C\|f\|_{\mathcal{L}^1} N^{-\frac{1}{2}},$$

with C a fixed constant.

Note that the exponent $s = 1/2$ is consistent with $p = 1$. The case a more general function $f \in H$ that does not have a summable expansion can be treated by the following result [6] which again holds for SP, OMP and RGA with the choice $\alpha_k := (1 - c/k)_+$.

Theorem 2.2. *If $f \in \mathcal{H}$, then for any $h \in \mathcal{L}^1$, we have*

$$\|f - f_N\|_{\mathcal{H}} \leq \|f - h\|_{\mathcal{H}} + C\|h\|_{\mathcal{L}^1} N^{-\frac{1}{2}}.$$

with C a fixed constant.

This result reveals that the accuracy of the greedy approximant is in some sense stable under perturbation, although the component selection process involved in the algorithm is unstable by nature.

An immediate consequence is that the greedy algorithm is convergent for any $f \in \mathcal{H}$ since we can approximate f to arbitrary accuracy by an $h \in \mathcal{L}^1$ (for example with a finite expansion in $\mathcal{D}$).

We can also use this result in order to identify more precisely the classes of functions which govern the approximation rate of the algorithm. Indeed, since the choice of $h \in \mathcal{L}^1$ is arbitrary, we have

$$\|f - f_N\|_{\mathcal{H}} \leq \inf_{h \in \mathcal{L}^1} \{\|f - h\|_{\mathcal{H}} + C\|h\|_{\mathcal{L}^1} N^{-1/2}\}.$$

The right hand side has the form of a so-called K-functional which is the central tool in the theory of interpolation space. Generally speaking, if X and Y are a pair of Banach function space, the corresponding K-functional is defined for all $f \in X + Y$ and $t > 0$ by

$$K(f,t) = K(f,t,X,Y) := \inf_{g \in X, h \in Y, g+h=f} \{\|g\|_X + t\|h\|_Y\}.$$

One then defines interpolation space by growth conditions on $K(f,t)$. In particular we say that $f \in [X,Y]_{\theta,\infty}$ (with $0 < \theta < 1$) if and only if there is a constant C such that for all $t > 0$,

$$K(f,t) \le Ct^\theta.$$

We refer to [8, 7] for general introductions on interpolation spaces. In our present setting, we see that

$$\|f - f_N\|_{\mathcal{H}} \le K(f, CN^{-\frac{1}{2}}, \mathcal{H}, \mathcal{L}^1),$$

and we therefore obtain

$$f \in [c\mathcal{H}, \mathcal{L}^1]_{\theta,\infty} \Rightarrow \|f - f_N\|_{\mathcal{H}} \le CN^{-s}, \quad s = \frac{\theta}{2}.$$

This result is consistent with the particular case of an orthonormal basis since in this case $\mathcal{H} \sim \ell^2(\mathcal{D})$ and $\mathcal{L}^1 \sim \ell^2(\mathcal{D})$ so that $[c\mathcal{H}, \mathcal{L}^1]_{\theta,\infty} \sim [\ell^2, \ell^1]_{\theta,\infty}$ which is known to coincide with the space $w\ell^p$ with $\frac{1}{p} = \theta + \frac{(1-\theta)}{2}$. We therefore recover the fact that $\|f - f_N\|_{\mathcal{H}} \le CN^{-s}$ when $(c_g)_{g \in \mathcal{D}} \in w\ell^p$ with $\frac{1}{p} = \frac{1}{2} + s$. For a more general dictionary, if we are able characterize the space $\mathcal{L}^1$ by some smoothness condition in $\mathcal{H}$, then $[c\mathcal{H}, \mathcal{L}^1]_{\theta,\infty}$ will correspond to some intermediate smoothness condition.

2.4 Open questions and related topics

The above results show that greedy algorithms have the convergence rate N^{-s} with $0 \le s \le \frac{1}{2}$ when f has a moderately concentrated expansion in $\mathcal{D}$.

At the other end, one might ask how the algorithm behaves when f has a highly concentrated expansion, i.e. $f = \sum_{g \in \mathcal{D}} c_g g$ with $(c_g)_{g \in \mathcal{D}} \in \ell^p$ for some $p < 1$. The limit case $p = 0$ of a finitely supported expansion corresponds to the *sparse recovery problem*: from the data of f can we recover its exact finite expansion by a fast algorithm?

For a general dictionary, it was proved in [17] that $(c_g)_{g \in \mathcal{D}} \in \ell^p$ with $p < 1$ implies the existence of a sequence f_N of N-terms approximant which converge towards f with the optimal rate N^{-s} with $s = \frac{1}{p} - 12$. However SP, OMP and RGA may fail to converge faster than $N^{-\frac{1}{2}}$. They may also fail to solve the sparse recovery problem.

On the other hand we know that SP, OMP and PGA are successful in the special case where $\mathcal{D}$ is an orthonormal basis. A natural question is therefore to understand the general conditions on a $\mathcal{D}$ under which the convergence of greedy algorithms might fully benefit of such concentration properties, similar to the case of an orthonormal basis. Important progress has been recently made in this direction, in relation with the topic of *compressed sensing*. We refer in particular to [18] in which it is proved that OMP succeeds with high probability in the sparse recovery problem for randomly generated dictionaries.

Other open questions concern the PGA algorithm for which it was proved in [17] that $f \in \mathcal{L}_1$ implies that

$$\|f - f_N\| \leq CN^{-\frac{1}{6}}.$$

This rate was improved to $N^{-\frac{11}{62}}$ in [20], but on the other hand it was shown [21] that for a particular dictionary there exists $f \in \mathcal{L}_1$ such that

$$\|f - f_N\| \geq cN^{-0.27}.$$

The exact best rate N^{-s} achievable for a general dictionary and $f \in \mathcal{L}^1$ is still unknown, but we already see that PGA is sub-optimal in comparison to SP, OMP and RGA, and an interesting issue is to understand which conditions should be imposed on the dictionary in order to recover an optimal rate of convergence.

3 Adaptive triangulations

We turn to the problem of designing optimally adapted triangulations for a function f defined on a bidimensional polygonal domain Ω. For the sake of simplicity, we shall only consider approximation by piecewise affine function. We therefore introduce the non-linear space Σ_N consisting of all functions which are piecewise affine on a triangulation of at most N triangles, i.e.

$$\Sigma_N := \cup_{\#(\mathcal{T}) \leq N} V_{\mathcal{T}},$$

where

$$V_{\mathcal{T}} := \{v \text{ s.t. } v_{|T} \in \Pi_1, \ T \in \mathcal{T}\}$$

is the piecewise affine space associated to $\mathcal{T}$. Given a norm X of interest, we define the best approximation error

$$\sigma_N(f)_X := \inf_{g \in \Sigma_N} \|f - g\|_X.$$

In the following we shall only consider $X = L^p$. In order to assess the performance of greedy algorithms, we first need discuss the fundamental question of approximation theory: what properties of f govern the rate of approximation in the sense that $\sigma_N(f)_X \leq CN^{-r}$ for all N with some given $r > 0$? As we recall below, the answer varies strongly when we consider uniform triangulations, adaptive triangulations with isotropy constraints or unconstrained adaptive triangulations.

3.1 From uniform to adaptive isotropic triangulations

Let us recall in a nutshell some approximation theory available for piecewise affine functions. If we first limitate our choice to families of *uniform* triangulations $(\mathcal{T}_h)_{h>0}$ where $h := \max\{\text{diam}(T), T \in \mathcal{T}_h\}$, classical finite element theory states that for all $s \leq 2$

$$f \in W^{s,p}(\Omega) \Rightarrow \inf_{g \in V_h} \|f - g\|_{L^p} \lesssim h^s |f|_{W^{s,p}},$$

where V_h is the space associated to $\mathcal{T}_h$. Since the triangulation is uniform, we have

$$N := \#(\mathcal{T}_h) \lesssim h^{-2},$$

and therefore, the approximation error satisfies

$$\sigma_N^{\text{unif}}(f)_{L^p} \lesssim |f|_{W^{s,p}} N^{-\frac{s}{2}}. \tag{3.1}$$

Consider next adaptive partitions with the restriction that all triangles are *isotropic* in the sense that

$$|T| \geq C(\text{diam}(T))^2,$$

for some fixed C independent of T. Combining Sobolev imbedding together with Deny-Lions theorem, we have the local polynomial approximation estimate

$$\min_{\pi \in \Pi_1} \|f - \pi\|_{L^p(T)} \leq C_0 |f|_{W^{s,q}(T)},$$

for $s \leq 2$ and $\frac{1}{q} = \frac{1}{p} + \frac{s}{2}$. The constant C_0 is invariant by dilation and therefore this estimate is uniform over all isotropic triangles. Assume now that we can construct adaptive isotropic triangulations $\mathcal{T}_N$ with $N := \#(\mathcal{T}_N)$ which *equidistribute* the local error in the sense that for some prescribed $\varepsilon > 0$

$$c\varepsilon \min_{\pi \in \Pi_1} \|f - \pi\|_{L^p(T)} \leq \varepsilon, \tag{3.2}$$

with $c > 0$ a fixed constant independent of T and N. Then defining f_N on each $T \in \mathcal{T}_N$ as the local minimizer of the L^p approximation error, we have on the one hand

$$\|f - f_N\|_{L^p} \leq N^{1/p}\varepsilon,$$

and on the other hand when $f \in W^{s,q}(\Omega)$

$$N(c\varepsilon)^q \leq \sum_{T \in \mathcal{T}_N} \|f - f_N\|_{L^p(T)}^q \leq \sum_{T \in \mathcal{T}_N} (C_0|f|_{W^{s,q}(T)})^q \leq C_0^q|f|_{W^{s,q}}^q.$$

Combining both, one obtains the approximation estimate

$$\sigma_N^{\mathrm{iso}}(f)_{L^p} \leq C|f|_{W^{s,q}}N^{-\frac{s}{2}}, \tag{3.3}$$

with $C = C_0/c$, which shows that the same rate as in (3.1) is now
obtained from a weaker smoothness condition since $q < p$ (note that
q could be less than 1 in which case the Sobolev space $W^{s,q}$ should
be replaced by the Besov space $B_{q,q}^s$). This type of result is classical in
non-linear approximation and also occurs when we consider best N-term
approximation in a wavelet basis.

The principle of error equidistribution suggests a simple *greedy algorithm* to build an adaptive isotropic triangulation for a given f:

1. Start from a coarse triangulation $\mathcal{T}_{N_0}$.

2. Given $\mathcal{T}_k$ split the triangle T which maximizes the local error $\|f -
   f_k\|_{L^p(T)}$ into four equal subtriangles using the mid-points of each
   side of T, as illustrated in Figure 3.1 left. This produces $\mathcal{T}_{k+3}$.

3. Stop when a prescribed error or number of triangle is met.

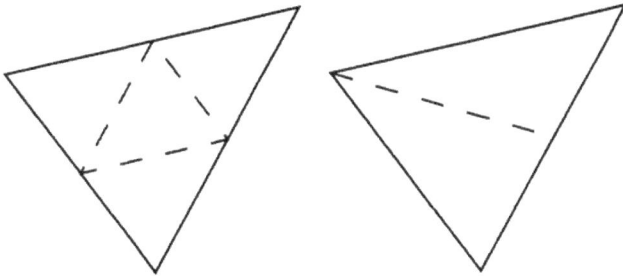

Figure 3.1   Isotropic quad-split (left) and anisotropic bisection (right).

Such an algorithm cannot exactly equilibrate the error in the sense of
(3.2) and therefore does not lead to the optimal estimate (3.3). However,
it is known that it satisfies

$$\|f - f_N\|_{L^p} \lesssim |f|_{W^{s,q}}N^{-\frac{s}{2}},$$

for all $s \leq q$ and q such that $\frac{1}{q} < \frac{1}{p} + \frac{s}{2}$ which is not far from the optimal
estimate (see [13] for this type of results expressed in a wavelet setting).

3.2   Approximation theory of anisotropic triangulations

One of the motivation for using anisotropic triangulations is when the function f has discontinuities along curved edges. As an intuitive example, consider $f = \chi_S$ with $S \subset \Omega$ is a set with smooth boundary ∂S. Then, adaptive isotropic triangulation has convergence rate

$$\sigma_N^{\mathrm{iso}}(f)_{L^p} \sim N^{-\frac{1}{p}},$$

which reflects the fact that the edge ∂S can resolved by a layer of N triangles which has width $\mathcal{O}(N^{-1})$. Allowing anisotropic triangles that are refined in the normal direction to the edge and aligned with the tangential direction, one can improve this width to $\mathcal{O}(N^{-2})$, which leads to the better rate

$$\sigma_N(f)_{L^p} \lesssim N^{-\frac{2}{p}}.$$

Similar improvement holds for more general functions which are piecewise C^m with discontinuities along piecewise C^n. However, it is not known if the available algorithms for data driven generation of anisotropic triangulations do produce approximation which satisfy the above rate.

   These observations have motivated the development of alternate strategies for the approximation of anisotropic features which are based on thresholding in appropriate dictionaries, such as curvelets [11], bandlets [22], edgeprints [4] or edge adapted multiscale transforms [1].

   For smooth functions, optimal approximation estimates for anisotropic triangulations can be derived from the following heuristic argument: if the hessian matrix $H(x) = D^2 f(x)$ is constant over a triangle T, we optimize the shape of T by imposing the same aspect ratio as the ellipsoid associated to $|H|$. If h_1 and h_2 are the height of T in the directions associated to the eigenvalues λ_1 and λ_2, we thus have

$$|T| \sim h_1 h_2,$$

and

$$h_1^2 \lambda_1 \sim h_2^2 \lambda_2.$$

Since the local approximation error is controlled in the uniform norm by

$$\inf_{\pi \in \Pi_1} \|f - \pi\|_{L^\infty(T)} \leq \lambda_1 h_1^2 + \lambda_2 h_2^2 \sim |T| \sqrt{\lambda_1 \lambda_2}.$$

it follows that the local L^p approximation error satisfies the estimate

$$\inf_{\pi \in \Pi_1} \|f - \pi\|_{L^p(T)} \leq C_1 |T|^{\frac{1}{q}} \sqrt{\det(|H|)},$$

with $\frac{1}{q} = 1 + \frac{1}{p}$. We again assume that we can construct anisotropic triangulations $\mathcal{T}_N$ with $N := \#(\mathcal{T}_N)$ which equidistribute the local error in the sense of (3.2), and that in addition the triangles are optimized in the above sense. Introducing the quantity

$$E(f) := \| \sqrt{\det(|H|)} \|_{L^q},$$

we now obtain

$$N(c\varepsilon)^q \le \sum_{T \in \mathcal{T}_N} \| f - f_N \|_{L^p(T)}^q \le C_1^q \sum_{T \in \mathcal{T}_N} |T| [\sqrt{\det(|H|)}]^q \le (C_1 E(f))^q$$

and therefore

$$\sigma_N(f)_{L^p} \le C N^{-1} E(f), \qquad (3.4)$$

with $C = C_1/c$. This estimate quantifies the improvement over isotropic triangulations, since in (3.3) the rate N^{-1} appeared with constant proportional to $|f|_{W^{2,q}} := \| D^2 f \|_{L^q}$ which might be much larger than $E(f)$. However it is wrong as such since $E(f)$ can theoretically vanish while the approximation error remains large (consider for instance f depending only on a single variable). A slightly different formulation which allows a rigorous analysis was proposed in [12]. In this formulation $|H| = |D^2 f|$ is replaced by $|H| := |D^2 f| + \varepsilon I$, avoiding that $E(f)$ vanishes (see also [2, 3]). The estimate (3.4) is then correct but holds for $N \ge N(\varepsilon, f)$ large enough. This limitation is unavoidable and reflects the fact that enough resolution is needed so that the Hessian can be viewed as locally constant over each optimized triangle.

3.3 A greedy algorithm for anisotropic triangulations

From an algorithmic perspective, the above heuristic observations have led to concrete algorithms which consist in designing the triangulation in such a way that each triangle is locally isotropic with respect to the metric associated to the absolute value Hessian. We refer in particular to [9, 10] where this program is executed using Delaunay mesh generation techniques.

While these algorithms fastly produce anisotropic meshes which are well adapted to the approximated function, they have two intrinsic limitations:

1. They are based on the evaluation of the hessian $D^2 f$, and therefore do not in principle apply to non-smooth functions or noisy data.

2. They are non-hierarchical: for $N > M$, the triangulation $\mathcal{T}_N$ is not a refinement of $\mathcal{T}_M$.

The following greedy algorithm, introduced in [14], constitutes an alternative approach which circumvents these limitations:

1. Start from an coarse triangulation $\mathcal{T}_{N_0}$.

2. Given $\mathcal{T}_k$ bisect the triangle T which maximizes the local error $\| f - f_k \|_{L^p(T)}$ from one of its vertex to the opposite mid-point, as illustrated

in Figure 3.1 right. The vertex is selected, between the three options so to minimize the local projection error for the new triangulation $\mathcal{T}_{k+1}$.

3. Stop when a prescribed error or number of triangle is met.

This type of adaptive bisection tends to generate anisotropic triangles which are well taylored to the geometry of curved singularities, such as edges in images or cliffs in terrain elevation data. As an example, Figure 3.2 displays the result of the algorithm after 512 steps, when applied (with the L^2 norm as a measure of local error) to the function $f(x) = y(x^2 + y^2) + \tanh(100(\sin(5y) - 2x))$ which has a sharp transition along the curve of equation $\sin(5y) = 2x$. Our algorithm behaves very well on this type of example in the sense that it develops anisotropic triangles along the transition curve.

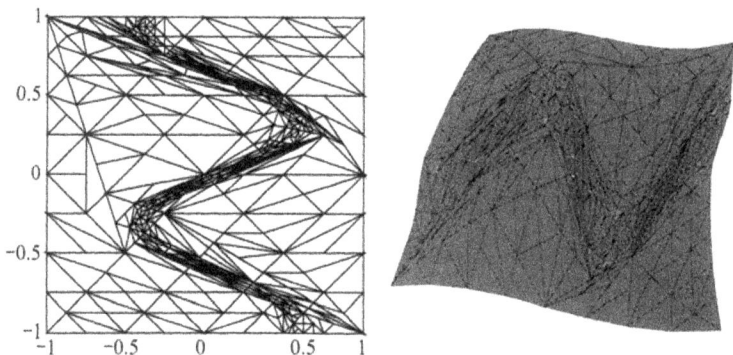

Figure 3.2   Triangulation (left) and approximation (right).

3.4   Convergence properties of the algorithm

Beside the apparently good behaviour of the algorithm how close are we from an optimal triangulation? Since we have limited our choice to a restricted family, there is in general no hope that the greedy algorithm can exactly produce the optimal one. In practice, we would be satisfied if we could show that the L^p error between the function and its approximation decays with the number of triangles at a rate which is similar to the optimal one expressed by (3.4). In particular, we want to understand if the triangles produced by the greedy algorithm tend to have the optimal aspect ratio in the sense that they are isotropic with respect to the metric induced by the hessian.

Let us indicate two results established in [14] which give some partial answer to this question, in the case where the greedy algorithm is applied with the L^2 norm as a measure of the local error. The first one describes the behaviour of the algorithm in a region where the Hessian is locally constant and positive, and shows that almost all triangles tend to have an optimal aspect ratio.

Theorem 3.1. *Let f is a quadratic function with a positive definite hessian $H = D^2 f$. If we apply the greedy bisection algorithm to this function with an initial triangulation T_{N_0}, then for all $\varepsilon > 0$ and for N large enough, there is at least $(1 - \varepsilon)N$ triangles in T_N which satisfy*

$$\{x \, ; \, \langle H(x-x_T), (x-x_T) \rangle \leq ch_N^2 \} \subset T \subset \{x \, ; \, \langle H(x-x_T), (x-x_T) \rangle \leq h_N^2 \},$$

where x_T is the barycenter of T, h_T a size parameter which only depends on N and $c < 1$ an absolute constant.

The second result describes the behaviour of the algorithm more globally when the function is C^2 and strictly convex, and shows that it is in accordance with (3.4).

Theorem 3.2. *Let f be a C^2 function that satisfies $D^2 f(x) \geq \alpha I$ for all $x \in \Omega$ with $\alpha > 0$ independent of x. If we apply the greedy bisection algorithm for this function with an initial triangulation T_{N_0}, then for N large enough we have*

$$\|f - f_N\|_{L^2} \leq C \| \sqrt{\det(D^2 f)} \|_{L^q} N^{-1}$$

with $\frac{1}{q} = 1 + \frac{1}{2}$ (i.e. $q = 2/3$) and C an absolute constant.

References

[1] F. Arandiga, A. Cohen, R. Donat, N. Dyn and B. Matei, *Approximation of piecewise smooth functions and images by edge-adapted (ENO-EA) nonlinear multiresolution transforms*, ACHA 24, 225–250, 2008.

[2] Y. Babenko, *Exact asymptotics of the uniform error of interpolation by multilinear splines*, to appear in J. Approximation Theory.

[3] V. Babenko, Y. Babenko, A. Ligun and A. Shumeiko, *On asymptotic behavior of the optimal linear spline interpolation error of C^2 functions*, East J. Approximation 12, 71–101, 2006.

[4] R. Baraniuk, H. Choi, J. Romberg and M. Wakin, *Wavelet-domain approximation and compression of piecewise smooth images*, IEEE Transactions on Image Processing, 2006.

[5] A. Barron, *Universal approximation bounds for superposition of n sigmoidal functions*, IEEE Trans. Inf. Theory **39**, 930–945, 1993.

[6] A. Barron, A. Cohen, W. Dahmen and R. DeVore, *Approximation and learning by greedy algorithms*, Annals of Statistics 36, 64–94, 2008.

[7] C. Bennett and R. Sharpley, *Interpolation of Operators*, in Pure and Applied Mathematics, Academic Press, N.Y., 1988.

[8] J. Bergh and J. Löfström, *Interpolation spaces*, Springer Verlag, Berlin, 1976.

[9] H. Bourouchaki, P.-L. George, F. Hecht, P. Laug and E. Saltel, *Delaunay mesh generation governed by metric specification. I. algorithms*, Finite Elem. Anal. Des. 25, 61–83, 1997.

[10] H. Bourouchaki, P.-L. George and B. Mohammadi, *Delaunay mesh generation governed by metric specification. II. applications*, Finite Elem. Anal. Des. 25, 85–109, 1997.

[11] E. Candes and D. L. Donoho, *Curvelets and curvilinear integrals*, J. Approx. Theory. 113, 59–90, 2000.

[12] L. Chen, P. Sun and J. Xu, *Optimal anisotropic meshes for minimizing interpolation error in L_p-norm*, Math of Comp., 2003.

[13] A. Cohen, W. Dahmen, I. Daubechies and R. DeVore, *Tree-structured approximation and optimal encoding*, App. Comp. Harm. Anal. **11**, 192–226, 2001.

[14] A. Cohen, N. Dyn, F. Hecht and J.-M. Mirebeau, *Adaptive multiresolution analysis based on anisotropic triangulations*, preprint Laboratoire J.-L. Lions, 2008.

[15] A. Cohen and J.-M. Mirebeau, *Greedy bisection generates optimally adapted triangulations*, preprint Laboratoire J.-L. Lions, 2008.

[16] R. DeVore, *Nonlinear approximation*, Acta Numerica, 51-150, 1998.

[17] R. DeVore and V. Temlyakov, *Some remarks on greedy algorithms*, Advances in Computational Mathematics **5**, 173–187, 1996.

[18] A. C. Gilbert and J. A. Tropp, *Signal recovery from partial information via orthogonal matching pursuit*, to appear in IEEE Trans. Inf. Theory.

[19] L. K. Jones, *A simple lemma on greedy approximation in Hilbert spaces and convergence rates for projection pursuit regression and neural network training*, Ann. Stat. **20**, 608–613, 1992.

[20] S. V. Konyagin and V. N. Temlyakov, *Rate of convergence of Pure greedy Algorithm*, East J. Approx. **5**, 493–499, 1999.

[21] E. D. Livshitz and V. N. Temlyakov *Two lower estimates in greedy approximation*, Constr. Approx. **19**, 509–524, 2003.

[22] E. Le Pennec and S. Mallat, *Bandelet image approximation and compression*, Multiscale Model. Simul. 4, 992–1039, 2005.

[23] V. Temlyakov, *Nonlinear methods of approximation* Journal of FOCM, **3**, 33–107, 2003.

[24] V. Temlyakov, *Greedy algorithms*, to appear in Acta Numerica.

The Contribution of Wavelets in Multifractal Analysis

Stéphane Jaffard[*]   Patrice Abry[†]   Stéphane G. Roux[*]

Béatrice Vedel[†*]   Herwig Wendt[*]

Abstract

We show how wavelet techniques allow to derive irregularity properties of functions on two particular examples: Lacunary Fourier series and some Gaussian random processes. Then, we work out a general derivation of the multifractal formalism in the sequence setting, and derive some of its properties.

The purpose of multifractal analysis is twofold: on the mathematical side, it allows to determine the size of the sets of points where a function has a given Hölder regularity; on the signal processing side, it yields new collections of parameters associated to the signal considered and which can be used for classification, model selection, or for parameter selection inside a parametric setting. The main advances in the subject came from a better understanding of the interactions between these two motivations. The seminal ideas of this subject were introduced by N. Kolmogorov in the years 1940, in the study of turbulence. Though they could have been used in other contexts at that time, they remained confined to this specific subject up to the mid 1980's. One reason is that the "scaling function" identified by Kolmogorov as a key-tool in the study of turbulence, was not clearly interpreted as a function-space index. Therefore, the subject could no benefit from the important advances performed from the 1950s to the 1980s in real analysis and function space theory. The situation changed completely in the mid 1980s for several reasons:

- The interpretation of the scaling function as a description of the statistical repartition of the pointwise Hölder singularities of the

---

[*]Address: Laboratoire d'Analyse et de Mathématiques Appliquées, UMR 8050 du CNRS, Université Paris Est, 61 Avenue du Général de Gaulle, 94010 Créteil Cedex, France.

[†]Address: CNRS UMR 5672 Laboratoire de Physique, ENS de Lyon, 46, allée d'Italie, F-69364 Lyon cedex, France.

signal by G. Parisi and U. Frisch supplied new motivations that
were no more specific to turbulence analysis. As a consequence,
these general methods were applied in many other settings.

• The wavelet formulation of the scaling function supplied ways to
  rewrite it that were fitted to modern signal processing; indeed, they
  were numerically more stable, they led to alternative, more robust
  definitions of the scaling function, and they allowed a mathematical
  analysis of these methods.

Our purpose in this paper is to describe these developments, and
to give an introduction to the recent research topics in this area. We
will also mention several open questions. It is partly a review based
on [1, 19, 20, 21, 23, 26], but it also contains original results. Recent
applications of these techniques in signal and image processing can be
found in [2, 3, 27, 41, 42].

1   Kolmogorov's scaling law and function spaces

Let us start by a short and partial description of the seminal work of
Kolmogorov in fully developed turbulence. The streamwise component
of turbulent flow velocity spatial field exhibits very irregular fluctua-
tions over a large range of scales, whose statistical moments furthermore
behave, within the so-called inertial scale range, like power laws with
respect to the scale h; this velocity measured at a given point is therefore a function of time only, which we denote by $v(t)$. This power-law
behavior is written

$$\int |v(t+h) - v(t)|^p dt \quad \sim \quad h^{\eta(p)}. \qquad (1.1)$$

This statement means that the function $\eta(p)$ can be determined as a
limit when $h \to 0$ on a log-log plot; it is called the scaling function of the
velocity v. Characterization and understanding of the observed scaling
properties play a central role in the theoretical description of turbulence,
and Kolmogorov in 1941 expected a linear scaling function for turbulent
flows: $\eta(p) = p/3$. This prediction has been refined by Obukhov and
Kolmogorov in 1962 who predicted a (quadratic) non-linear behavior of
the scaling exponents. The non-linear behavior of $\eta(p)$ was confirmed
by various experimental results and other models have been proposed
leading to different scaling functions $\eta(p)$. Let us now give the function
space interpretation of this initial scaling function. This is done with the
help of the spaces Lip(s, L^p) defined as follows. (We give definitions

in the d-dimensional setting, since we will deal with several variable functions later on.)

Definition 1. *Let* $s \in (0, 1)$, *and* $p \in [1, \infty]$; $f \in Lip(s, L^p(\mathbb{R}^d))$ *if* $f \in L^p$ *and* $\exists C > 0$ *such that* $\forall h > 0$,

$$\| f(x + h) - f(x) \|_{L^p} \leq Ch^s. \tag{1.2}$$

Note that, if s is larger than 1, one uses differences of higher order in (1.2). It follows from this definition that, if $\eta(p) < p$,

$$\eta(p) = \sup\{s : f \in \text{Lip}(s/p, L^p(\mathbb{R}^d))\}. \tag{1.3}$$

Remark: The condition $\eta(p) < p$ has to hold because this interpretation is valid only if the smoothness exponent s in Definition 1 is less than 1; otherwise, the scaling function should be defined using higher order differences. The spaces $\text{Lip}(s, L^p)$ are defined only for $p \geq 1$, however, in applications, one also considers values of the scaling function for $p < 1$; therefore, one would like to extend the function space interpretation of the scaling function to smaller values of p in a proper mathematical way. Finally, the scaling law in (1.1) is much more precise than what is given by the function space interpretation (1.3): Indeed, the order of magnitude of the integral $\int |v(t + h) - v(t)|^p dt$ might oscillate indefinitely between two power laws $h^{\eta_1(p)}$ and $h^{\eta_2(p)}$ when $h \to 0$. One way to avoid this problem is to define the scaling function directly by the function space interpretation, i.e. by (1.3). Thus, if

$$S(f, p, h) = \int_{\mathbb{R}^d} |f(x + h) - f(x)|^p dx, \qquad \text{then}$$

$$\eta(p) = \liminf_{h \to 0} \frac{\log(S(f, p, h))}{\log(h)}. \tag{1.4}$$

Using (1.4) as a definition has the further advantage of allowing to drop the assumption $\eta(p) < p$. In practice, the scaling function is determined by plotting, for each p, $\log(S(f, p, h))$ as a function of $\log(h)$: it can be sharply estimated only if one obtains a plot which is close to a straight line on a sufficient number of scales. This means in particular that, if in (1.4) the liminf is not a real limit, then one cannot expect to determine the scaling function. Note also that (1.3) and (1.4) are not changed if one modified the right hand-side of (1.2) by logarithmic corrections, i.e. if it were of the form $|h|^s |\log(1/h)|^\beta$ for instance. This remark allows to give alternative interpretations of the scaling function based on other function spaces, and to extend it to values of p that lie between 0 and 1. Let us recall the definition of other families of alternative function spaces currently used.

Definition 2. *Let $s \geq 0$ and $p \geq 1$. A function f belongs to the Sobolev space $L^{p,s}(\mathbb{R}^d)$ if $f \in L^p$ and if $(Id - \Delta)^{s/2} f \in L^p$, where the operator $(Id - \Delta)^{s/2}$ is defined as follows: $g = (Id - \Delta)^{s/2} f$ means that $\hat{g}(\xi) = (1 + |\xi|^2)^{s/2} \hat{f}(\xi)$ (the function $(1 + |\xi|^2)^{s/2}$ being C^∞ with polynomial increase, $(1 + |\xi|^2)^{s/2} \hat{f}(\xi)$ is well defined if f is a tempered distribution).*

This definition amounts to say that f and its fractional derivatives of order at most s belong to L^p. Let us now recall the definition of Besov spaces. If s is large enough, Besov spaces can be defined by conditions on the finite differences $\Delta_h^M f$ which are defined as follows.

Let $f : \mathbb{R}^d \to \mathbb{R}$ and $h \in \mathbb{R}^d$. The first order difference of f is

$$(\Delta_h^1 f)(x) = f(x + h) - f(x).$$

If $n > 1$, the differences of order n are defined recursively by

$$(\Delta_h^n f)(x) = (\Delta_h^{n-1} f)(x + h) - (\Delta_h^{n-1} f)(x).$$

Definition 3. *Let p, q and s be such that $0 < p \leq +\infty$, $0 < q \leq +\infty$ and $s > d(\frac{1}{p} - 1)_+$; then $f \in B_p^{s,q}(\mathbb{R}^d)$ if $f \in L^p$ and if, for $M > s$,*

$$\int_{|h| \leq 1} |h|^{-sq} \|(\Delta_h^M f)\|_p^q \frac{dh}{|h|^d} \leq C. \tag{1.5}$$

If $s \geq 0$ and $p \geq 1$, then the following embeddings hold

$$B_p^{s,1} \hookrightarrow L^{p,s} \hookrightarrow B_p^{s,\infty} \quad \text{and} \quad B_p^{s,1} \hookrightarrow \mathrm{Lip}(s, L^p(\mathbb{R}^d)) \hookrightarrow B_p^{s,\infty}$$

and, if $s \geq 0$, $p > 0$ and $0 < q_1 < q_2$, then $\forall \epsilon > 0$,

$$B_p^{s+\epsilon,\infty} \hookrightarrow B_p^{s,q_1} \hookrightarrow B_p^{s,q_2} \hookrightarrow B_p^{s,\infty}.$$

Thus $B_p^{s,q}$ is "very close" to $L^{p,s}$ and $\mathrm{Lip}(\alpha, L^p)$. (Recall also that $B_\infty^{\alpha,\infty} = C^\alpha(\mathbb{R}^d)$.) In fact, all the previous families of function spaces coincide "up to logarithmic corrections" in the sense mentioned previously. More precisely, they are equivalent in the following sense:

Definition 4. *Let A_p^s and B_p^s denote two families of function spaces. They are equivalent families of function spaces in the range (p_1, p_2) if*

$$\forall p \in (p_1, p_2), \quad \forall \epsilon > 0 \qquad A_p^{s+\epsilon} \subset B_p^s \subset A_p^{s-\epsilon}.$$

The scaling function of a function f in the scale A_p^s is

$$\eta^A(p) = \sup\{s : f \in A_p^{s/p}\}. \tag{1.6}$$

Thus, the scaling function associated with two equivalent families coincide. This abstract setting will also be useful in Section 5 where we will consider families of function spaces defined for sequences instead of functions. The embeddings between $\text{Lip}(s, L^p)$, Sobolev and Besov spaces imply that they are equivalent families of function spaces in the range $(1, \infty)$; and, when q varies, Besov spaces are equivalent families in the range $(0, \infty)$. Therefore, in (1.3), one can replace $\text{Lip}(\alpha, L^p)$ spaces by Sobolev or Besov spaces without altering the definition of the scaling function. The notion of equivalent families of function spaces has a practical motivation: Since the scaling function of a signal is numerically determined by a slope in a log-log plot, one cannot numerically draw a difference between two scales that are equivalent.

Since Besov spaces are defined for $p > 0$, the definition of the scaling function with Besov spaces allows to extend it in a natural way to all values of $p > 0$:

$$\forall p > 0, \quad \forall q > 0, \qquad \eta(p) = \sup\{s : f \in B_q^{s/p,p}\}. \qquad (1.7)$$

It follows that, for $p > 0$, any of the equivalent definitions of Besov or Sobolev spaces which have been found can be used in the determination of the scaling function of a signal. Wavelet characterizations are the ones that are now preferred in practice. We will see how the extend $\eta(p)$ to negative values of p (this will require a detour via wavelet-based formulas for scaling functions).

Let us mention another problem posed by function-space modeling when applied to real-life signals: Data are always available on a finite length; therefore modeling should use function spaces defined on an interval (or a domain, in several dimensions). This leads to several complications, especially when dealing with wavelets, and is not really relevant, when boundary phenomena are not of interest. Therefore, one uses wavelet bases and function spaces defined on $\mathbb{R}$ or $\mathbb{R}^d$.

2   Pointwise regularity

2.1   Hölder exponents

Pointwise regularity is a way to quantify, using a positive parameter α, the fact that the graph of a function may be more or less "ruguous" at a point x_0.

Definition 5. *Let α be a nonnegative real number, and $x_0 \in \mathbb{R}^d$; a function $f : \mathbb{R}^d \to \mathbb{R}$ belongs to $C^\alpha(x_0)$ if there exists $C > 0$, $\delta > 0$ and a polynomial P satisfying $\deg(P) < \alpha$ such that*

$$\text{if } |x - x_0| \leq \delta, \qquad |f(x) - P(x - x_0)| \leq C|x - x_0|^\alpha. \qquad (2.1)$$

The Hölder exponent of f at x_0 is

$$h_f(x_0) = \sup\{\alpha : \ f \ \text{is} \ C^\alpha(x_0)\}.$$

The polynomial P is unique; the constant term of P is necessarily $f(x_0)$; P is called the Taylor expansion of f at x_0 of order α. The notion of Hölder exponent is adapted to functions whose regularity changes abruptly from point to point. When it is not the case, the more stable notion of *local Hölder exponent* can be used:

$$H_f(x_0) = \inf\{\alpha : \ \exists \delta > 0, \ f \in C^\alpha([x_0 - \delta, x_0 + \delta])\},$$

see [28]. However, up to now, this notion has had no impact on multifractal analysis: Indeed, most functions with a nonconstant pointwise Hölder exponent which have been considered have a constant local Hölder exponent (in which case, the local Hölder exponent does not allow to draw distinctions between different local behaviors).

Note that (2.1) implies that f is bounded in a neighborhood of x_0; therefore, the Hölder exponent can be defined only for locally bounded functions. The Hölder exponent is defined point by point and describes the local regularity variations of f. Some functions have a constant Hölder exponent; they are called *monohölder functions*. It is the case of the Weierstrass functions that we will consider in Section 3. We will see that it is also the case for Brownian motion, and Fractional Brownian Motions. Such functions display a "very regular irregularity". On the opposite, "multifractal functions" have a very irregular Hölder exponent which cannot be estimated numerically point by point.

2.2   Other notions of pointwise regularity

The notion of pointwise Hölder regularity is pertinent only if applied to locally bounded functions. We will derive pointwise irregularity results for solutions of PDEs where the natural function space setting is L^p or Sobolev spaces. In such cases, one has to use the following extension of pointwise smoothness, which was introduced by Calderón and Zygmund in 1961, see [7].

Definition 6. *Let $B(x_0, r)$ denote the open ball centered at x_0 and of radius r; let $p \in [1, +\infty)$ and $\alpha > -d/p$. Let f be a tempered distribution on $\mathbb{R}^d$; f belongs to $T_\alpha^p(x_0)$ if it coincides with an L^p function in a ball $B(x_0, R)$ for an $R > 0$, and if there exist $C > 0$ and a polynomial P of degree less than α such that*

$$\forall r \leq R, \quad \left(\frac{1}{r^d} \int_{B(x_0, r)} |f(x) - P(x - x_0)|^p dx\right)^{1/p} \leq C r^\alpha. \quad (2.2)$$

The p-exponent of f at x_0 is

$$h_f^p(x_0) = \sup\{\alpha : f \in T_\alpha^p(x_0)\}.$$

Note that the Hölder exponent corresponds to the case $p = +\infty$, and the condition on the degree of P implies its uniqueness. This definition is particularly useful when dealing with functions which are not locally bounded: it is a natural substitute for pointwise Hölder regularity when functions in L_{loc}^p are considered. In particular, the p-exponent can take values down to $-d/p$, and therefore it allows to model behaviors which are locally of the form $1/|x - x_0|^\alpha$ for $\alpha < d/p$. For example, this is relevant in fully developed turbulence where singularities of negative Hölder exponent corresponding to thin vorticity filaments can be observed, see [4]. We will use this notion in Section 3, in order to derive everywhere irregularity results for solutions of PDEs.

Pointwise Hölder regularity can also be considered in the setting of measures.

Definition 7. *Let $x_0 \in \mathbb{R}^d$ and let $\alpha \geq 0$. A probability measure μ defined on $\mathbb{R}^d$ belongs to $C^\alpha(x_0)$ if there exists a constant $C > 0$ such that, in a neighborhood of x_0,*

$$\mu(B(x_0, r)) \leq Cr^\alpha.$$

Let x_0 belong to the support of μ: then the lower Hölder exponent of μ at x_0 is

$$h_\mu(x_0) = \sup\{\alpha : \mu \in C^\alpha(x_0)\}.$$

The upper Hölder exponent of μ at x_0 is

$$\tilde{h}_\mu(x_0) = \inf\{\alpha \quad \text{such that, for } r \text{ small enough,} \quad \mu(B(x_0, r)) \geq r^\alpha\}.$$

Note that the Hölder exponent of a measure is sometimes called the *local dimension*. We will need to deduce the Hölder exponent at every point from discrete quantities, which, in practice, will be indexed by the dyadic cubes.

Definition 8. *Let $j \in \mathbb{Z}$; a dyadic cube of scale j is of the form*

$$\lambda = \left[\frac{k_1}{2^j}, \frac{k_1 + 1}{2^j}\right) \times \cdots \times \left[\frac{k_d}{2^j}, \frac{k_d + 1}{2^j}\right), \qquad (2.3)$$

where $k = (k_1, \cdots k_d) \in \mathbb{Z}^d$.

Each point $x_0 \in \mathbb{R}^d$ is contained in a unique dyadic cube of scale j, denoted by $\lambda_j(x_0)$.

The cube $3\lambda_j(x_0)$ is the cube of same center as $\lambda_j(x_0)$ and three times wider; i.e., if $\lambda_j(x_0)$ is given by (2.3), then it is the cube

$$3\lambda_j(x_0) = \left[\frac{k_1 - 1}{2^j}, \frac{k_1 + 2}{2^j}\right) \times \cdots \times \left[\frac{k_d - 1}{2^j}, \frac{k_d + 2}{2^j}\right).$$

Using dyadic cubes in analysis has two advantages: For a fixed j, they form a partition of $\mathbb{R}^d$, and they are naturally endowed with a tree structure which is inherited from the notion of inclusion: A dyadic cube of scale j is exactly composed of 2^j dyadic "children" of scale $j + 1$. This tree structure will play a key-role in the notion of *wavelet leader* in Section 4.3, where pointwise Hölder regularity of functions will be characterized in terms of quantities defined on the dyadic cubes. It is also the case for the pointwise exponents of measures we introduced: One immediately checks that they can be derived from the knowledge of the quantities $\omega_\lambda = \mu[3\lambda]$:

- Let μ be a nonnegative measure defined on $\mathbb{R}^d$. Then

$$\forall x_0, \qquad h_\mu(x_0) = \liminf_{j \to +\infty} \left(\frac{\log\left(\mu[3\lambda_j(x_0)]\right)}{\log(2^{-j})} \right). \qquad (2.4)$$

- Similarly,

$$\forall x_0, \qquad \tilde{h}_\mu(x_0) = \limsup_{j \to +\infty} \left(\frac{\log\left(\mu[3\lambda_j(x_0)]\right)}{\log(2^{-j})} \right). \qquad (2.5)$$

We will now study some Gaussian processes and lacunary Fourier series, which show why wavelet-type techniques yield irregularity results on technically easy examples.

2.3 Brownian motion and related noncentered stochastic processes

Brownian motion (defined on $\mathbb{R}^+$) is (up to a multiplicative constant) the only random process with independent and stationary increments which has continuous sample paths; i.e., if $y > x$, $B(y) - B(x)$ is independent of the $B(t)$ for $t \leq x$, and has the same law as $B(y - x)$. P. Lévy and Z. Ciesielski obtained a remarkable decomposition of the Brownian motion restricted on $[0, 1]$. Let $\Lambda(x)$ be the "hat function" defined by

$$\Lambda(x) = \begin{cases} x & \text{if } x \in [0, 1/2], \\ 1 - x & \text{if } x \in [1/2, 1], \\ 0 & \text{else.} \end{cases}$$

The Brownian motion restricted on $[0, 1]$ can be written

$$B(x) = \chi_0 x + \sum_{j \geq 0} \sum_{k=0}^{2^j - 1} 2^{-j/2} \chi_{j,k} \Lambda(2^j x - k) \qquad (2.6)$$

where χ_0 and the $\chi_{j,k}$ are Gaussian independent identically distributed (IID) random variables of variance 1. The set of functions 1, x, and the

$$\Lambda(2^j x - k), \quad j \geq 0, \ k = 0, \cdots, 2^j - 1, \qquad (2.7)$$

which appear in (2.6), is called the *Schauder basis* on the interval $[0, 1]$. Note that (2.7) has the same algorithmic form as wavelet bases (see Section 4.1): indeed, the Schauder basis is obtained by taking the primitives of the Haar basis (and renormalizing them correctly). Therefore it is not surprising that the technique we introduce now in order to estimate the pointwise smoothness of the Brownian motion anticipates the wavelet techniques of Section 4.3.

Definition 9. *Let* $f : \mathbb{R}^+ \to \mathbb{R}$ *be a given continuous function. The decentered Brownian motion of expectation f is the stochastic process*

$$X(x) = f(x) + B(x).$$

This denomination is justified by the fact that $\forall x$, $\mathbb{E}(X(x)) = f(x)$.

Theorem 1. *Let f be an arbitrary continuous function on $[0, 1]$. With probability 1, the sample paths of X satisfy*

$$\forall x_0 \in [0, 1] \quad \limsup_{x \to x_0} \frac{|X(x) - X(x_0)|}{\sqrt{|x - x_0|}} > 0.$$

Therefore the Hölder exponent of X satisfies

$$a.s. \qquad \forall x \qquad h_X(x) = \inf\left(h_f(x), \frac{1}{2}\right). \tag{2.8}$$

Remark: This theorem can be interpreted as an example of a generic result in the sense of *prevalence*. Such results, where one proves that a property holds "almost surely" in a given function space E require the proof of results of this type, which hold for the sum of an arbitrary function $f \in E$ and of a stochastic process whose sample path almost surely belong to E, see [12, 16, 17]. Here, the corresponding prevalent result is that the Hölder exponent of almost every continuous function is everywhere less than $1/2$. The prevalent notion of genericity offers an alternative which is often preferred to the previous notion of "quasi-sure" in the sense of Baire categories.

In order to prove Theorem 1, we will need two ancillary lemmas. The first one yields the explicit formula of the coefficients of a continuous function on the Schauder basis.

Lemma 1. *Let f be a continuous function on $[0, 1]$; then*

$$f(x) = f(0) + x(f(1) - f(0)) + \sum_{j \geq 0} \sum_{k=0}^{2^j - 1} C_{j,k} \Lambda(2^j x - k)$$

where

$$C_{j,k} = 2f\left(\frac{k + 1/2}{2^j}\right) - f\left(\frac{k}{2^j}\right) - f\left(\frac{k + 1}{2^j}\right).$$

We give the idea of the proof of this well known result: One checks by recursion on J that

$$P_J(f)(x) = f(0) + x(f(1) - f(0)) + \sum_{j=0}^{J} \sum_{k=0}^{2^j-1} C_{j,k} \Lambda(2^j x - k)$$

is the continuous piecewise linear function which coincides with f at the points $l/2^{J+1}$; the uniform convergence of $P_J(f)$ to f follows from the uniform continuity of f on $[0,1]$.

Lemma 2. *Let f be a continuous function on $[0,1]$. Let $k_j(x)$ be the integer k such that $x \in [\frac{k}{2^j}, \frac{k+1}{2^j}[$. If*

$$|f(x) - f(x_0)| \le C\sqrt{|x - x_0|}, \tag{2.9}$$

then its Schauder coefficients on $[0,1]$ satisfy

$$|C_{j,k_j(x_0)}| \le 4C2^{-j/2}.$$

Proof of Lemma 2: Using Lemma 1,

$$|C_{j,k_j(x_0)}| = \left| 2f\left(\frac{k_j(x_0) + 1/2}{2^j}\right) - f\left(\frac{k_j(x_0)}{2^j}\right) - f\left(\frac{k_j(x_0) + 1}{2^j}\right) \right|$$

$$\le 2\left| f\left(\frac{k_j(x_0) + 1/2}{2^j}\right) - f(x_0) \right| + \left| f\left(\frac{k_j(x_0)}{2^j}\right) - f(x_0) \right|$$

$$+ \left| f\left(\frac{k_j(x_0) + 1}{2^j}\right) - f(x_0) \right|,$$

which, using (2.9) is bounded by $4C2^{-j/2}$.

Proof of Theorem 1: Denote by $C_{j,k}$ the coefficients of X on the Schauder basis. We call a (C, j_0)-*slow point* a point x_0 where the sample path of $X(x)$ satisfies (2.9) for any x such that $|x - x_0| \le 2^{-j_0}$. If x_0 is such a point, using Lemma 2, for all couples $(j, k_j(x_0))$ such that $j \ge j_0$, we have

$$|C_{j,k_j(x_0)}| \le 4C2^{-j/2}. \tag{2.10}$$

Let $f_{j,k}$ denote the Schauder coefficients of f. Using (2.6), (2.10) can be rewritten

$$\forall j \ge j_0, \qquad |2^{-j/2}\chi_{j,k_j(x_0)} + f_{j,k_j(x_0)}| \le 4C2^{-j/2}.$$

For any j, k, let

$$p_{j,k} = \mathbb{P}\left(|2^{-j/2}\chi_{j,k} + f_{j,k}| \le 4C2^{-j/2} \right).$$

Since the $\chi_{j,k}$ are standard Gaussians,

$$p_{j,k} = \sqrt{\frac{2}{\pi}} \int_{-4C - 2^{j/2} f_{j,k}}^{4C - 2^{j/2} f_{j,k}} e^{-x^2/2} dx \leq 8C \sqrt{\frac{2}{\pi}} := p_C.$$

Let j_0 be given, $j \geq j_0$ and λ be a dyadic interval of length 2^{-2j}. If $l \leq 2j$, let $k_l(\lambda)$ denote the integer k such that $\lambda \in [\frac{k}{2^l}, \frac{k+1}{2^l}[$. Let E_λ be the event defined by:

$$\forall l \in \{j+1, \cdots, 2j\}, \qquad |C_{l,k_l(\lambda)}| \leq 4C 2^{-l/2}.$$

Since the Schauder coefficients of B are independent, the probability of E_λ satisfies $\mathbb{P}(E_\lambda) \leq (p_C)^j$. Since there are 2^{2j} such intervals λ, the probability that at least one of the events E_λ occurs is bounded by $(p_C)^j 2^{2j}$. Since $p_C \leq 8C\sqrt{\frac{2}{\pi}}$, we see that, if $C < \frac{1}{32}\sqrt{\frac{2}{\pi}}$, this probability tends to 0, and therefore a.s. there is no (C, j_0)-slow point in the interval $[0,1]$; the result is therefore true for any j_0, hence the first part of the theorem holds.

The second part of the theorem is a consequence of the following classical results:

- $h_{f+g}(x) \geq \inf(h_f(x), h_g(x))$;

- if $h_f(x) \neq h_g(x)$, then $h_{f+g}(x) = \inf(h_f(x), h_g(x))$;

- the Hölder exponent of the Brownian motion is everywhere $1/2$;

(2.8) follows for the points where the Hölder exponent of f differs from $1/2$. Else, $h_{f+B} \geq 1/2$, and the first part of the theorem implies that it is at most $1/2$.

An extension of Theorem 1 to the Fractional Brownian Motion will be proved in Section 4.4.

3 Lacunary Fourier series

Our second example of estimation of Hölder exponents is supplied by lacunary Fourier series. Informally, a lacunary Fourier series satisfies the following property: "Most" of its coefficients vanish. We will obtain a general result of irregularity based on the *Gabor-wavelet transform*, and we will apply it to the case of multidimensional nonharmonic Fourier series. Examples of applications to everywhere irregularity results for solutions of PDEs will then be derived.

3.1 A pointwise irregularity criterium

We will use the following notation: If λ, $x \in \mathbb{R}^d$, $\lambda \cdot x$ denotes the usual scalar product of λ and x.

Definition 10. *Let $\phi : \mathbb{R}^d \longrightarrow \mathbb{R}$ be a function in the Schwartz class such that $\hat{\phi}(\xi)$ is supported in the unit ball centered at 0 and such that $\hat{\phi}(0) = 1$. The Gabor-wavelet transform of a function or a tempered distribution f defined on $\mathbb{R}^d$ is defined by*

$$d(a, b, \lambda) = \frac{1}{a^d} \int_{\mathbb{R}^d} f(x) e^{-i\lambda \cdot (x-b)} \phi\left(\frac{x-b}{a}\right) dx. \qquad (3.1)$$

Note that, if we pick λ of the form $\lambda = \lambda_0/a$ where λ_0 is fixed and does not belong to $supp(\phi)$, then this definition boils down to the continuous wavelet transform in several dimensions; the purpose of adding the extra factor $e^{-i\lambda \cdot (x-b)}$ is to bring an additional frequency shift which will prove useful.

Proposition 1. *Let f be a tempered distribution; let $p \in (1, +\infty]$ and assume that f belongs to L^p. Let $\alpha > -d/p$; if $f \in T_\alpha^p(x_0)$, then there exists $C' > 0$, which depends only on ϕ and α such that*

$$\forall a > 0, \quad \forall \lambda : \quad |\lambda| \geq \frac{1}{a}, \quad |d(a, b, \lambda)| \leq CC' a^\alpha \left(1 + \frac{|x_0 - b|}{a}\right)^{\alpha + d/p},$$
$$(3.2)$$

where C is the $T_\alpha^p(x_0)$ constant that appears in (2.2).

Proof of Proposition 1: If $w_{a,\lambda}(x) = a^{-d} e^{-i\lambda \cdot x} \phi(x/a)$, then

$$\widehat{w_{a,\lambda}}(\xi) = \hat{\phi}(a(\xi + \lambda)),$$

so that, as soon as $|\lambda| > 1/a$, $\widehat{w_{a,\lambda}}$ and all its derivatives vanishes at 0. It follows that

$$d(a, b, \lambda) = \frac{1}{a^d} \int_{\mathbb{R}^d} (f(x) - P(x - x_0)) e^{i\lambda \cdot x} \phi\left(\frac{x-b}{a}\right) dx.$$

For $n \geq 0$, let $B_n = B(b, 2^n a)$, $\Delta_n = B_{n+1} - B_n$ and $\Delta_0 = B_0$. We split $d(a, b, \lambda)$ as a sum of integrals I_n over Δ_n. Let q denote the conjugate exponent of p; by Hölder's inequality,

$$\forall n \geq 0, \quad |I_n| \leq \frac{1}{a^d} \| f(x) - P(x - x_0) \|_{L^p(B_{n+1})} \left\| \phi\left(\frac{x-b}{a}\right) \right\|_{L^q(\Delta_n)}.$$

Since $B_{n+1} \subset B(x_0, |x_0 - b| + 2^{n+1}a)$, and since ϕ has fast decay, $\forall D$ large enough,

$$|I_n| \leq \frac{CC'(D)}{a^d} (|x_0 - b| + 2^{n+1}a)^{\alpha + d/p} a^{d/q} (2^{-Dn})^{1/q}.$$

Therefore

$$|d(a, b, \lambda)| \leq CC'(D)a^{-d/p} \sum_{n=0}^{\infty} (|x_0 - b| + 2^{n+1}a)^{\alpha+d/p} 2^{-Dn/q}$$
$$\leq CC' \left(|x_0 - b|^{\alpha+d/p} a^{-d/p} + a^{\alpha} \right),$$

hence Proposition 1 holds.

3.2   Application to nonharmonic Fourier series

Let $(\lambda_n)_{n \in \mathbb{N}}$ be a sequence of points in $\mathbb{R}^d$. We will consider series of the form

$$f(x) = \sum_{n \in \mathbb{N}} a_n e^{i\lambda_n \cdot x}, \tag{3.3}$$

where $(a_n)_{n \in \mathbb{N}}$ is a sequence of complex numbers with, at most, polynomial increase. Of course, we can (and will) assume that the λ_n are distinct. Note that we do not assume that the λ_n are integers; one usually refers to (3.3) as *nonharmonic Fourier series*.

Definition 11. *Let (λ_n) be a sequence in $\mathbb{R}^d$. The gap sequence associated with (λ_n) is the sequence (θ_n) defined by*

$$\theta_n = \inf_{m \neq n} |\lambda_n - \lambda_m|.$$

The sequence (λ_n) is separated if $\inf_n \theta_n > 0$.

Note that θ_n is the distance between λ_n and its closest neighbour. We will always assume in this section that (λ_n) is separated and (a_n) increases at most polynomially, which implies the convergence of (3.3) in the space of tempered distributions.

Proposition 2. *Let f be given by (3.3) and let x_0 be a given point of $\mathbb{R}^d$, $p > 1$, $\alpha > -d/p$ and assume that f belongs to L^p in a neighborhood of x_0. If $f \in T_{\alpha}^p(x_0)$, then there exists C' which depends only on α such that*

$$\forall n \in \mathbb{N} \quad if \ |\lambda_n| \geq \theta_n, \quad then \quad |a_n| \leq \frac{CC'}{\theta_n^{\alpha}}, \tag{3.4}$$

where C is the constant that appears in (2.2).

Remark: This is indeed an everywhere irregularity result: Let

$$H = \sup\{\alpha : \ (3.4) \ \text{holds}\};$$

Proposition 2 implies that the p-exponent of f is everywhere smaller than H. (Since the p-exponent is larger than the Hölder exponent, if

f is locally bounded it also implies that the Hölder exponent of f is everywhere smaller than H.)

Proof of Proposition 2: Let us estimate the Gabor-wavelet transform of f at particular points, and for a function ϕ such that $\hat{\phi}(\xi)$ is radial, supported in the unit ball centered at 0, and such that $\hat{\phi}(0) = 1$. Let

$$D_m = d\left(\frac{1}{\theta_m}, x_0, \lambda_m\right). \tag{3.5}$$

On one hand,

$$D_m = (\theta_m)^d \int \left(\sum_n a_n e^{i(\lambda_n - \lambda_m)\cdot x} \phi(\theta_m(x - x_0))\right) dx$$

$$= \sum_n a_n \hat{\phi}\left(\frac{\lambda_m - \lambda_n}{\theta_n}\right) e^{i(\lambda_n - \lambda_m)\cdot x_0}; \tag{3.6}$$

since $\hat{\phi}$ vanishes outside of $B(0, 1)$, the definition of θ_n implies that $\hat{\phi}\left(\frac{\lambda_m - \lambda_n}{\theta_n}\right) = \delta_{n,m}$, so that $D_m = a_m$. On the other hand, if $f \in T_\alpha^p(x_0)$, then Proposition 1 implies that, for any m such that $|\lambda_m| \geq \theta_m$, $|D_m| \leq C\theta_m^{-\alpha}$; Proposition 2 follows.

3.3 Everywhere irregularity of solutions of Schrödinger's equation

As a consequence of the previous results, let us show that the solutions of a simple linear PDE display a remarkable property of everywhere irregularity if the initial condition is not smooth. We consider the one-dimensional Schrödinger equation

$$i\frac{\partial \psi}{\partial t} = -\frac{\partial^2 \psi}{\partial x^2}, \qquad \text{for} \quad (x, t) \in \mathbb{R} \times \mathbb{R} \tag{3.7}$$

with initial condition:

$$\psi(x, 0) = \psi_0(x) = \sum_{n \in \mathbb{Z}} a_n e^{inx}. \tag{3.8}$$

The general solution of (3.7) can be written

$$\psi(x, t) = \sum_{n \in \mathbb{Z}} a_n e^{inx} e^{-in^2 t}. \tag{3.9}$$

Note that (3.9) is of the form $\sum_{n \in \mathbb{Z}} a_n e^{i \lambda_n \cdot X}$ where $\lambda_n = (n, -n^2)$, and $X = (x, t)$, so that the gap sequence θ_n associated with λ_n satisfies

$$\forall n \in \mathbb{Z}, \qquad \theta_n \geq |n| + 1. \qquad (3.10)$$

Let $p > 1$ and $\alpha \geq -2/p$. It follows from Proposition 2 that

$$\text{if} \quad \psi \in T_\alpha^p(x_0, t_0), \quad \text{then} \quad \forall n, \quad |a_n| \leq \frac{C}{(|n| + 1)^\alpha}. \qquad (3.11)$$

But, if $|a_n| \leq C(|n| + 1)^{-\alpha}$, then $\psi_0(x)$ belongs to the periodic Sobolev space H^s, as soon as $s < \alpha - 1/2$. One can also consider the trace $\tilde{\psi}_{x_0}$ of ψ at a given point x_0, as a function of t, i.e. formally, $\tilde{\psi}_{x_0}(t) = \psi(x_0, t)$. The solution is still a one-dimensional lacunary Fourier series with $\lambda_n = n^2$. We obtain that, if $\tilde{\psi}_{x_0}(t) \in T_\alpha^1(t_0)$, then $|a_n| \leq C/n^\alpha$. Hence the following corollary holds.

Corollary 1. *Let $s > -5/2$, let $\psi(x, t)$ be a solution of (3.7), and assume that $\psi_0 \notin H^s$. Then*

$$\forall \alpha > s + 1/2, \quad \forall (x_0, t_0), \quad \forall p > 1, \qquad \psi \notin T_\alpha^p(x_0, t_0).$$

Furthermore, as regards irregularity in the time direction,

$$\forall \alpha > s + 1/2, \quad \forall (x_0, t_0), \quad \forall p > 1, \qquad \psi_{x_0} \notin T_\alpha^p(x_0, t_0).$$

In particular, if (3.7) has a bounded solution, $\forall \alpha > s + 1/2, \forall (x_0, t_0)$, $\psi_{x_0} \notin C^\alpha(t_0)$.

The fact that solutions of the Schrödinger equation are everywhere irregular somehow means that their graph is a fractal; in a similar spirit, the fractal properties of the graph of the fundamental solution of (3.7) have been investigated by K. Oskolkov, see [36] and references therein.

4 Wavelets, function spaces and Hölder regularity

Orthonormal wavelet bases are a privileged tool to study multifractal functions for several reasons. A first one, exposed in this section, is that classical function spaces (such as Besov or Sobolev spaces) can be characterized by conditions bearing on the wavelet coefficients, see Section 4.2; furthermore, pointwise regularity can also be characterized by simple local decay conditions on the wavelet coefficients, see Section 4.3. Another reason concerns the wavelet formulation of the multifractal formalism, which leads to the construction of new function spaces, directly defined by conditions on the wavelet coefficients. We will just recall some properties of orthonormal and biorthogonal wavelet bases that will be useful in the following. We refer the reader for instance to [10, 11, 29, 34] for detailed expositions of this subject.

4.1   Orthonormal and biorthogonal wavelet bases

Orthonormal wavelet bases are of the following form: There exists a function $\varphi(x)$ and $2^d - 1$ functions $\psi^{(i)}$ with the following properties: The functions $\varphi(x - k)$ $(k \in \mathbb{Z}^d)$ and the $2^{dj/2}\psi^{(i)}(2^j x - k)$ $(k \in \mathbb{Z}^d,$ $j \in \mathbb{Z})$ form an orthonormal basis of $L^2(\mathbb{R}^d)$. This basis is r-smooth if φ and the $\psi^{(i)}$ are C^r and if the $\partial^\alpha \varphi$, and the $\partial^\alpha \varphi \psi^{(i)}$, for $|\alpha| \leq r$, have fast decay.

Therefore, $\forall f \in L^2$,

$$f(x) = \sum_{k \in \mathbb{Z}^d} C_k \varphi(x - k) + \sum_{j=0}^{\infty} \sum_{k \in \mathbb{Z}^d} \sum_i c^i_{j,k} \psi^{(i)}(2^j x - k); \qquad (4.1)$$

the $c^i_{j,k}$ are the wavelet coefficients of f

$$c^i_{j,k} = 2^{dj} \int_{\mathbb{R}^d} f(x) \psi^{(i)}(2^j x - k) dx, \qquad (4.2)$$

and

$$C_k = \int_{\mathbb{R}^d} f(x) \varphi(x - k) dx. \qquad (4.3)$$

Remark: In (4.1), we do not choose the L^2 normalisation for the wavelets, but rather an L^∞ normalisation which is better fitted to the study of Hölder regularity. The L^1 normalisation of (4.2) follows accordingly.

Note that (4.2) and (4.3) make sense even if f does not belong to L^2; indeed, if one uses smooth enough wavelets, these formulas can be interpreted as a duality product between smooth functions (the wavelets) and distributions. We will see the examples of Sobolev and Besov spaces.

We will also need decompositions on *biorthogonal wavelet bases*, which are a useful extension of orthonormal wavelet bases. A *Riesz basis* of an Hilbert space H is a collection of vectors (e_n) such that the finite linear expansions $\sum_{n=1}^N a_n e_n$ are dense in H and

$$\exists C, C' > 0: \ \forall N, \quad \forall a_n, \quad C \sum_{n=1}^N |a_n|^2 \leq \left\| \sum_{n=1}^N a_n e_n \right\|_H^2 \leq C' \sum_{n=1}^N |a_n|^2.$$

Two collections of functions (e_n) and (f_n) form *biorthogonal bases* if each collection is a Riesz basis, and if $\langle e_n | f_m \rangle = \delta_{n,m}$. When such is the case, any element $f \in H$ can be written

$$f = \sum_{n=1}^{\infty} \langle f | f_n \rangle e_n. \tag{4.4}$$

Biorthogonal wavelet bases are couples of Riesz bases of L^2 which are, of the form: on one side,

$$\varphi(x - k), \quad (k \in \mathbb{Z}^d) \quad \text{and} \quad 2^{dj/2} \psi^{(i)}(2^j x - k), \quad (k \in \mathbb{Z}^d, \ j \in \mathbb{Z})$$

and, on the other side,

$$\tilde{\varphi}(x - k) \quad (k \in \mathbb{Z}^d) \quad \text{and} \quad 2^{dj/2} \tilde{\psi}^{(i)}(2^j x - k), \quad (k \in \mathbb{Z}^d, \ j \in \mathbb{Z}).$$

Therefore, $\forall f \in L^2$,

$$f(x) = \sum_{k \in \mathbb{Z}^d} C_k \varphi(x - k) + \sum_{j=0}^{\infty} \sum_{k \in \mathbb{Z}^d} \sum_i c_{j,k}^i \psi^{(i)}(2^j x - k); \tag{4.5}$$

where

$$c_{j,k}^i = 2^{dj} \int_{\mathbb{R}^d} f(x) \tilde{\psi}^{(i)}(2^j x - k) dx \quad \text{and} \quad C_k = \int_{\mathbb{R}^d} f(x) \tilde{\varphi}(x - k) dx. \tag{4.6}$$

We will see that biorthogonal wavelet bases are particularly well adapted to the decomposition of the Fractional Brownian Motion: indeed, well chosen biorthogonal wavelet bases allow to decorrelate the wavelet coefficients of these processes (the wavelet coefficients become independent random variables), and therefore greatly simplifies their analysis.

We will use more compact notations for indexing wavelets. Instead of using the three indices (i, j, k), we will use dyadic cubes. Since i takes $2^d - 1$ values, we can assume that it takes values in $\{0, 1\}^d - (0, \cdots, 0)$: we introduce:

- $\lambda \ (= \lambda(i, j, k)) \ = \dfrac{k}{2^j} + \dfrac{i}{2^{j+1}} + \left[0, \dfrac{1}{2^{j+1}}\right)^d$,

- $c_\lambda = c_{j,k}^i$,

- $\psi_\lambda(x) = \psi^{(i)}(2^j x - k)$.

The wavelet ψ_λ is essentially localized near the cube λ; more precisely, when the wavelets are compactly supported

$$\exists C > 0 \quad \text{such that} \quad \forall i, j, k, \qquad supp\,(\psi_\lambda) \subset C \,\lambda$$

(where $C\,\lambda$ denotes the cube of same center as λ and C times wider). Finally, Λ_j will denote the set of dyadic cubes λ which index a wavelet of scale j, i.e. wavelets of the form $\psi_\lambda(x) = \psi^{(i)}(2^j x - k)$ (note that Λ_j is a subset of the dyadic cubes of side 2^{j+1}). We take for norm on $\mathbb{R}^d$

$$\text{if } x = (x_1, \cdots, x_d), \quad |x| = \sup_{i=1,\cdots,d} |x_i|;$$

so that the diameter of a dyadic cube of side 2^{-j} is exactly 2^{-j}.

Among the families of wavelet bases that exist, two will be particularly useful for us:

- Lemarié-Meyer wavelets, such that φ and $\psi^{(i)}$ both belong to the Schwartz class;

- Daubechies wavelets, such that the functions φ and $\psi^{(i)}$ can be chosen arbitrarily smooth and with compact support.

If the wavelets are r-smooth, they have a corresponding number of vanishing moments, see [34]:

$$\text{If } |\alpha| < r, \quad \text{then} \quad \int_{\mathbb{R}^d} \psi^{(i)}(x) x^\alpha dx = 0.$$

Therefore, if the wavelets are in the Schwartz class, all their moments vanish.

4.2   Wavelets and function spaces

A remarkable property of wavelet bases is that they supply bases not only in the L^2 setting, but also for most function spaces that are used in analysis. When considering wavelet characterization of function spaces, a first natural question is to understand in which sense the wavelet series of a function, or of a distribution converges, i.e. in which sense wavelets are *bases* of the corresponding space; before giving the two standard definitions of bases (depending whether E is separable or not, see [38]), we need to recall the notions of quasi-norm and quasi-Banach space.

Definition 12. *Let E be a vector space. A quasi-norm on E is a non-negative function satisfying*

$$\begin{aligned} &\exists C, \ \forall x, y \in E, \ \| x + y \| \le C(\| x \| + \| y \|), \\ &\forall \lambda \in \mathbb{R}, \ \forall x \in E, \ \| \lambda x \| = \| x \|, \\ &\forall x \in E, \quad\quad\quad \| x \| = 0 \implies x = 0. \end{aligned}$$

A quasi-Banach space is a vector space endowed with a quasi-norm, and which is complete for the corresponding topology.

Besov spaces for $p < 1$ or $q < 1$ are typical example of quasi-Banach spaces, indeed (1.5), or equivalently (4.9), only define quasi-norms if p or q are less than 1.

Definition 13. *Let E be a Banach, or a quasi-Banach space. A sequence e_n is a basis of E if the following condition holds: For any element f in E, there exists a unique sequence c_n such that the partial sums $\sum_{n \leq N} c_n e_n$ converge to f in E. It is an unconditional basis if furthermore*

$$\exists C > 0, \ \forall \epsilon_n \ \text{such that} \ |\epsilon_n| \leq 1, \ \forall c_n, \ \| \sum c_n \epsilon_n e_n \|_E \leq C \, \| \sum c_n e_n \|_E \, .$$
(4.7)

If the space E is not separable (it is the case for instance of Besov spaces when p or q is infinite, or C^α spaces, since $C^\alpha = B^\infty_{\alpha,\infty}$) then, of course, it cannot have a basis in the previous sense. In this case, the following weaker notion often applies.

Definition 14. *Assume that E is the dual of a separable space F; a sequence $e_n \in E$ is a weak* basis of E if, $\forall f \in E$, there exists a unique sequence c_n such that the partial sums $\sum_{n \leq N} c_n e_n$ converge to f in the weak* topology. It is unconditional if furthermore (4.7) holds.*

Let F be either the dual of E (in the basis setting of Definition 13) or a predual of E (in the weak* basis setting); we will furthermore always assume in the following that,

$$\text{if } f = \sum c_n e_n, \quad \text{then there exist } g_n \in F \quad \text{such that } c_n = \langle f | g_n \rangle.$$
(4.8)

The g_n are called the biorthogonal system of the e_n; indeed, this notion extends the previous definition of biorthogonality in the non-Hilbert setting (in the Hilbert case where $E = F$, (4.8) boils down to (4.4)). Note that if E is a Banach space, if $F = E^*$ and if the e_n form a basis according to Definition 13, then (4.8) is automatically verified, see [38]; it is also verified if the e_n are a wavelet basis, in which case $g_n = e_n$ for L^2 orthonomal wavelet bases (or g_n is another wavelet basis in the wavelet biorthogonal case). Note that, for wavelets, the L^2 biorthogonal system is also the biorthogonal system for the (E, F) duality; indeed, by uniqueness if $\mathcal{S}_0$ is dense in either E or F, then the $(\mathcal{S}_0, \mathcal{S}_0')$ duality, the (L^2, L^2) duality and the (E, F) duality coincide for finite linear combinations of wavelets; therefore (4.8) holds for all functions of E by density, and the duality product $\langle f | g_n \rangle$ in (4.8) can be understood in any of the three settings. These considerations explain why the usual L^2 wavelet decomposition (4.1) also makes sense in other function space settings.

Examples of non-separable spaces for which wavelets are weak* bases include the Hölder spaces $C^s(\mathbb{R}^d)$, and, more generally, the Besov spaces $B_p^{s,q}$ with $p = +\infty$ or $q = +\infty$. We now give the wavelet characterizations of the spaces that will be useful for us. These characterizations supply equivalent norms or quasi-norms for the corresponding spaces, see [34].

Proposition 3. *Let ψ_λ be an r-smooth wavelet basis with $r > \sup(s, s + d(\frac{1}{p} - 1))$. Let $s > 0$ and $p, q \in (0, \infty]$. A function f belongs to the Besov space $B_p^{s,q}(\mathbb{R}^d)$ if and only if $(c_k) \in l^p$ and*

$$\sum_{j \in \mathbb{Z}} \left(\sum_{\lambda \in \Lambda_j} \left[2^{(s-d/p)j} |c_\lambda| \right]^p \right)^{q/p} \le C \qquad (4.9)$$

(using the usual convention for l^∞ when p or q in infinite).

A function f belongs to $L^{p,s}(\mathbb{R}^d)$ (for $1 < p < +\infty$) if and only if $(c_k) \in l^p$ and

$$\left(\sum_{\lambda \in \Lambda} |2^{sj} c_\lambda|^2 1_\lambda(x) \right)^{1/2} \in L^p(\mathbb{R}^d). \qquad (4.10)$$

Remark: When p or q is infinite, it may come as a surprise that non-separable spaces are characterized by expansion properties on a countable set of functions. However, Proposition 3 does **not** state that the partial sums of the wavelet series (4.1) converge in the corresponding space (which would indeed be in contradiction with nonseparability); it only yields a quantity which is equivalent to the Besov or Sobolev norm. Note however that, in the separable case, partial sums of (4.1) do converge in the corresponding space. Wavelets are a basis of Sobolev or Besov spaces when $p < \infty$ and $q < \infty$; else they are a weak* basis.

4.3 Wavelet characterizations of pointwise regularity

Pointwise Hölder regularity is characterized in terms of the following quantities.

Definition 15. *The wavelet leaders are*

$$d_\lambda = \sup_{\lambda' \subset 3\lambda} |c_{\lambda'}|. \qquad (4.11)$$

We note $d_j(x_0) = d_{\lambda_j(x_0)}$.

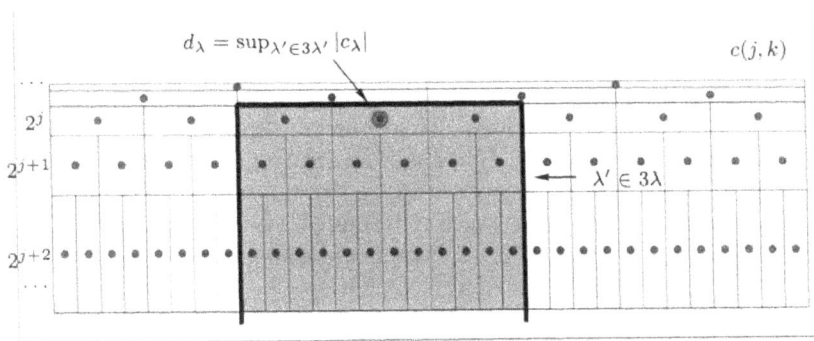

Figure 4.1 **Definition of wavelet Leaders.** The wavelet Leader d_λ (red circle) is defined as the largest wavelet coefficient $c_{\lambda'}$ (blue dots) within the time neighborhood 3λ (grey area) over all finer scales.

Figure 4.1 gives an illustration of the construction of the wavelet leaders in dimension 1.

If $f \in L^\infty$, then

$$|c_\lambda| \leq 2^{dj} \int_{\mathbb{R}^d} |f(x)||\psi_\lambda(x)|dx \leq C \sup_{x \in \mathbb{R}^d} |f(x)|,$$

so that the wavelet leaders are finite. The wavelet characterization of the Hölder exponent requires the following regularity hypothesis, which is slightly stronger than continuity.

Definition 16. *A function f is uniform Hölder if there exists $\epsilon > 0$ such that $f \in C^\epsilon(\mathbb{R}^d)$.*

The following theorem allows to characterize the pointwise regularity by a decay condition of the $d_j(x_0)$ when $j \to +\infty$, see [18] for the first statement of this result, and [20] for its reformulation in terms of wavelet leaders; see also [25] for similar results in the setting of general moduli of continuity.

Theorem 2. *Let $\alpha > 0$ and let ψ_λ be an orthonormal basis (or let $(\psi_\lambda, \tilde{\psi}_\lambda)$ be a couple of biorthogonal wavelet bases) with regularity $r > \alpha$. If f is $C^\alpha(x_0)$, then there exists $C > 0$ such that*

$$\forall j \geq 0, \quad d_j(x_0) \leq C2^{-\alpha j}. \tag{4.12}$$

Conversely, if (4.12) holds and if f is uniform Hölder, then there exist $C > 0$, $\delta > 0$ and a polynomial P satisfying $\deg(P) < \alpha$ such that

$$if \ |x - x_0| \leq \delta, \quad |f(x) - P(x - x_0)| \leq C|x - x_0|^\alpha \log\left(1/|x - x_0|\right). \tag{4.13}$$

Remark: Some uniform regularity is a necessary assumption in the converse part of Theorem 2: One can show that there exist bounded functions satisfying (4.12) for arbitrary large values of α, and whose Hölder exponent at x_0 vanishes. The reason of this phenomenon will be clarified in Section 4.3. Let f be a uniform Hölder function; since (4.13) only involves a logarithmic correction of the modulus of continuity, the regularity of f at x_0 is therefore determined by the decay rate of the $d_j(x_0)$ on a log-log plot. Hence a formula similar to (2.4) holds:

If f is a uniform Hölder function, then

$$h_f(x_0) = \liminf_{j \to +\infty} \left(\frac{\log(d_j(x_0))}{\log(2^{-j})} \right). \tag{4.14}$$

The Hölder exponent supplies a definition of pointwise regularity which can be difficult to handle or irrelevant; here are a few reasons:

- Mathematical results concerning multifractal analysis based on the Hölder exponent necessarily make the assumption that f is continuous, see Theorem 2; in many situations, one wishes to analyze discontinuous functions; an important case is natural images which, because of the *occlusion* phenomenon (one object is partially hidden by another), present discontinuities. There exists other fields where one even has to consider non-locally bounded functions (for instance in the study of fully developed turbulence, see [4]).

- A standard function space setting used for the mathematical study of images is supplied by the space of functions of bounded variations $BV(\mathbb{R}^2)$. Recall that $f \in BV$ if ∇f (defined in the sense of distributions) is a bounded measure. In dimension 1, BV functions are bounded (but may be discontinuous); in dimension 2, they can be nowhere locally bounded; however $BV(\mathbb{R}^2) \subset L^2$ and therefore the natural setting to perform the multifractal analysis of BV functions in dimension 2 is to use $T^2_\alpha(x_0)$ regularity instead of $C^\alpha(x_0)$ regularity, see [15].

- Let $\Omega \subset \mathbb{R}^d$ be a domain of $\mathbb{R}^d$ with a fractal boundary. A possible way to perform a multifractal analysis of Ω consists in associating to its characteristic function 1_Ω a pointwise regularity exponent. The Hölder exponent is clearly not the right tool since, in this case, it can take only two values: 0 on the boundary $\partial\Omega$ and $+\infty$ elsewhere.

Another reason, based on "stability" requirements, will be detailed in the following; they are related to the fact that the condition $f \in C^\alpha(x_0)$ is not invariant under simple pseudodifferential operators of order 0, and equivalently, cannot be characterized by conditions on the moduli of the wavelet coefficients of f.

These considerations motivated the use of $T_\alpha^p(x_0)$ regularity, introduced in Definition 6 as an alternative criterium of pointwise regularity (already considered in Section 3). One immediately checks that the $T_\alpha^p(x_0)$ regularity condition is weaker than Hölder regularity: If $f \in C^\alpha(x_0)$, then, $\forall p$, $f \in T_\alpha^p(x_0)$. The drawbacks of the $C^\alpha(x_0)$ criterium of smoothness that we listed above disappear when one considers this notion of regularity; for instance, the Hölder exponent of a characteristic function 1_Ω only takes the value 0 along the boundary of Ω; on the opposite, consider the "cusp domain" $\Omega \subset \mathbb{R}^2$ defined by the conditions

$$(x, y) \in \Omega \quad \text{if} \quad 0 \le |x| \le |y|^\beta, \qquad \text{for a } \beta \ge 1;$$

at the origin, the p-exponent of 1_Ω is $(\beta - 1)/p$ which can take any nonnegative value. Therefore, the p-exponent of a characteristic function 1_Ω can freely vary along the boundary of Ω, thus opening the way to a multifractal analysis of domains, see [24]. Furthermore, mathematical results concerning a multifractal analysis based on the p-exponent do not require any uniform regularity assumption, see [22].

Let us come back to the initial problem we mentioned, i.e. the instability of the $C^\alpha(x_0)$ condition. Indeed, the initial motivation of Calderón et Zygmund was to understand how pointwise regularity conditions are transformed in the resolution of elliptic PDEs, and they introduced the $T_u^p(x)$ spaces because the standard pseudodifferential operators of order 0 are not continuous on $C^\alpha(x_0)$, whereas, it is the case for the $T_u^p(x)$ spaces. We point out how this deficiency of the $C^\alpha(x_0)$ condition can be put into light. We consider the Hilbert transform, which is the simplest possible singular integral operator in dimension 1, and also plays a key-role in signal processing; it is the convolution with the principal value of $1/x$, i.e. is defined by

$$\mathcal{H}f(x) = \lim_{\epsilon \to 0} \frac{1}{\pi} \int_{I_\epsilon(x)} \frac{f(y)}{x - y} dy,$$

where $I_\epsilon(x) = (-\infty, x - \epsilon] \cup [x + \epsilon, +\infty)$. An immediate computation shows that

$$\mathcal{H}(1_{[a,b]})(x) = \log \left| \frac{x - b}{x - a} \right|. \tag{4.15}$$

Let now $(x_n)_{n \in \mathbb{N}}$ be a strictly decreasing sequence such that $\lim_{n \to \infty} x_n = 0$. We can pick a positive, strictly decreasing sequence a_n such that

$$f = \sum_{n=1}^{\infty} a_n 1_{[x_{n+1}, x_n]} \tag{4.16}$$

is arbitrarily smooth at 0. Nonetheless, (4.15) implies that

$$\mathcal{H}f(x) = \sum_{n=1}^{\infty} a_n \log \left| \frac{x - x_{n+1}}{x - x_n} \right|$$

$$= -a_1 \log |x - x_1| + \sum_{n=1}^{\infty} (a_n - a_{n+1}) \log |x - x_{n+1}|,$$

which is not locally bounded near the origin, and therefore cannot have any Hölder regularity there. Note that what we really used here is the fact that the Hilbert transform is not continuous on L^∞.

We can actually reinterpret the previous counterexample in a way that sheds some light on the necessity of the uniform Hölder assumption in Theorem 2. We will need the following lemma.

Lemma 3. *Let ψ be a wavelet generating an r-smooth orthonormal wavelet basis. Then the $\tilde{\psi}_{j,k} = \mathcal{H}(\psi_{j,k})$ also form an r-smooth orthonormal wavelet basis.*

We just sketch the proof: First, we recall that, if $f \in L^2$, then

$$\widehat{\mathcal{H}(f)}(\xi) = sgn(\xi)\hat{f}(\xi). \tag{4.17}$$

It follows that $\tilde{\psi} = \mathcal{H}(\psi)$ has the same number of vanishing moments as ψ. The smoothness of $\tilde{\psi}$ follows from the continuity of the Hilbert transform on the Hölder spaces $\dot{C}^\alpha$. The vanishing moments of ψ up to order r imply that $\tilde{\psi}$ and its derivatives up to order r decay like $(1 + |x|)^{-r-1}$. Note that (4.17) implies that the Hilbert transform is an L^2 isometry; therefore the functions $\mathcal{H}(\psi_{j,k})$ form an orthonormal basis of $L^2(\mathbb{R})$. Since the Hilbert transform is given by a convolution kernel, it commutes with translations, so that $\mathcal{H}(\psi_{j,k})$ is deduced from $\mathcal{H}(\psi_{j,0})$ by a translation of $k2^{-j}$. Because of the homogeneity of degree 0 of the function $sgn(\xi)$, it follows that the $\mathcal{H}(\psi_{j,0})$ are deduced from $\mathcal{H}(\psi)$ by a dyadic dilation. Therefore $(\mathcal{H}(\psi))_{j,k} = \mathcal{H}(\psi_{j,k})$; hence Lemma 3 holds.

Suppose now that a wavelet characterization of $C^\alpha(x_0)$ did exist; consider the function f defined by (4.16); its coefficients on the $\tilde{\psi}_{j,k}$ would satisfy this characterization. But

$$\langle f | \tilde{\psi}_{j,k} \rangle = \langle f | \mathcal{H}(\psi_{j,k}) \rangle = \langle \mathcal{H}(f) | \psi_{j,k} \rangle,$$

so that the criterium would be satisfied by the coefficients of $\mathcal{H}(f)$ on the $\psi_{j,k}$, which is absurd, since $\mathcal{H}(f)$ is not locally bounded in a neighborhood of x_0.

We now turn to the wavelet characterization of $T_\alpha^p(x_0)$. It is derived from the wavelet characterization of L^p, which is a particular case of (4.10), where $s = 0$: We obtain that $f \in L^p$ if

$$\left(\sum_{\lambda \in \Lambda} |2^{sj} c_\lambda|^2 1_\lambda(x) \right)^{1/2} \in L^p(\mathbb{R}^d). \qquad (4.18)$$

It is natural to expect that $T_\alpha^p(x_0)$ regularity will be characterized by a local condition bearing on the quantities involved in (4.18).

Definition 17. *Let ψ_λ be a given wavelet basis on $\mathbb{R}^d$. The local square function is*

$$S_{f,\lambda}(x) = \left(\sum_{\lambda' \subset 3\lambda} |c_{\lambda'}|^2 1_{\lambda'}(x) \right)^{1/2}.$$

The following theorem of [22] gives the wavelet characterization of $T_\alpha^p(x_0)$ (see also [24]).

Theorem 3. *Let $p \in (1, \infty)$ and $f \in L^p$. Let $\alpha > -d/p$ and assume that the wavelet basis used is r-smooth with $r > \sup(2\alpha, 2\alpha + 2d(\frac{1}{p} - 1))$; let*

$$d_\lambda^p = 2^{dj/p} \| S_{f,\lambda} \|_p .$$

If $f \in T_\alpha^p(x_0)$, then $\exists C \geq 0$ such that $\forall j \geq 0$,

$$d_{\lambda_j(x_0)}^p \leq C 2^{-\alpha j}. \qquad (4.19)$$

Conversely, if (4.19) holds and if $\alpha \notin \mathbb{N}$, then $f \in T_\alpha^p(x_0)$.

Remark: In contradistinction with Theorem 2, this result does not require a uniform regularity assumption. If $p = 2$, this characterization boils down to a local l^2 condition on the wavelet coefficients, since

$$d_\lambda^2 = \left(2^{dj} \sum_{\lambda' \subset 3\lambda_j(x_0)} 2^{-dj'} |c_{\lambda'}|^2 \right)^{1/2}. \qquad (4.20)$$

Theorem 3 can be given the following interpretation which is similar to (2.4) for Hölder exponents of measures and to (4.14) for Hölder exponents of functions:

Lemma 4. *Let $p \in (1, \infty)$ and $f \in L^p$. Then*

$$h_f^p(x_0) = \liminf_{j \to +\infty} \left(\frac{\log \left(d_{\lambda_j(x_0)}^p \right)}{\log(2^{-j})} \right). \qquad (4.21)$$

4.4 Application to decentered Fractional Brownian Motions

Fractional Brownian Motion (denoted by FBM) of index γ $(0 < \gamma < 1)$ is the only centered Gaussian random process $B^\gamma(x)$ satisfying

$$E(|B^\gamma(x) - B^\gamma(y)|^2) = |x - y|^{2\gamma}.$$

We will use the following important feature: FBM of index γ can be deduced from the Gaussian white noise by a fractional integration of order $\gamma + 1/2$. With probability 1, a sample path of FBM of order γ has everywhere the Hölder exponent γ.

Since the Gaussian white noise $N(x)$ has standard Gaussian I.I.D. coefficients on any orthonormal basis, using a wavelet basis, we obtain

$$N(x) = \sum_{j,k} \chi_{j,k} 2^{j/2} \psi(2^j x - k).$$

Let ψ be in the Schwartz class and

$$\hat{\psi}_\alpha(\xi) = \frac{1}{|\xi|^\alpha} \hat{\psi}(\xi) \qquad (4.22)$$

(ψ_α is the fractional integral of ψ of order α). If the wavelet ψ has enough vanishing moments, then ψ_α is a wavelet and one easily checks that the $2^{j/2}\psi_\alpha(2^j x - k)$ and the $2^{j/2}\psi_{-\alpha}(2^j x - k)$ form biorthogonal bases. The point of using these bases in order to analyze FBM is that, as a consequence of the previous remarks, the coefficients of FBM are decorrelated on it. More precisely,

$$B_\gamma(x) = \sum_{j=0}^{\infty} \sum_{k \in \mathbb{Z}} 2^{-\gamma j} \chi_{j,k} \, \psi_{\gamma+1/2}(2^j x - k) + R(x) \qquad (4.23)$$

where R is a C^∞ random process, and the $\xi_{j,k}$ are I.I.D. standard centered Gaussians. The following result extends Theorem 1 to the case of decentered FBMs.

Theorem 4. *Let f be an arbitrary L^2 function. Let*

$$X(x) = f(x) + B^\gamma(x).$$

With probability 1, the sample paths of X satisfy

$$\forall x_0 \in \mathbb{R}, \quad \limsup_{x \to x_0} \frac{|X(x) - X(x_0)|}{|x - x_0|^\gamma} > 0.$$

Therefore the Hölder exponent of X satisfies

$$a.s. \qquad \forall x \qquad h_X(x) = \inf\left(h_f(x), \gamma\right).$$

An illustration of this result is shown in Figure 4.2 for the theoretical and estimated spectrum of $X = f + B^\gamma$ with f a multifractal random walk (see [32] for its definition) and $\gamma = 0.7$.

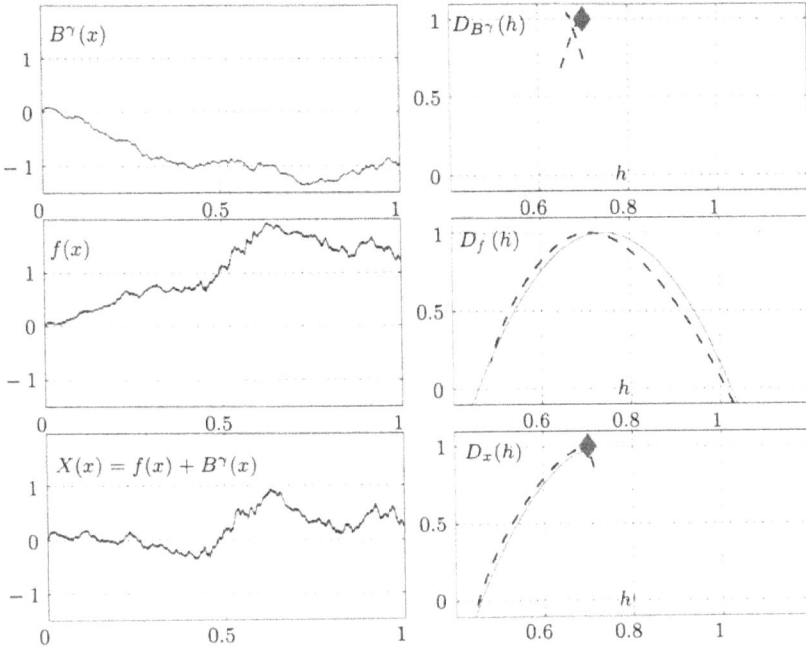

Figure 4.2 **Superimposition of functions.** Right column: Theoretical multifractal spectrum (red solid line) together with its Leader based estimation (dashed black line) for three different processes (left column). Top, fractional Brownian motion (FBM) with $\gamma = 0.7$; Middle, multifractal random walk (MRW) with scaling function $\zeta(p) = c_1 p + c_2 p^2 / 2$ ($c_1 = 0.72$ and $c_2 = -0.04$) [5]; Bottom additive superimposition of the two previous processes. The bottom right plot clearly shows that for additive superimposition the actual Hölder exponent at each x_0 actually corresponds to the minimum of each function, hence the Hölder exponents of MRW that are larger than $\gamma = 0.7$ are no longer present in FBM + MRW, as detected by the leader based estimation.

Remark: The prevalent implication of this theorem is that the Hölder exponent of almost every continuous function is everywhere at most γ. Since this is true for any $\gamma > 0$ we recover that the Hölder exponent of almost every continuous function vanishes everywhere, see [16].

Proof of Theorem 4: Let us denote by $C_{j,k}$ the coefficients of X on the wavelet basis generated by $\psi_{\gamma+1/2}$. We call a (C, j_0)-*slow point* a point x_0 where the sample path of $X(x)$ satisfies

$$|X(x) - X(x_0)| \leq C|x - x_0|^\gamma$$

for any x such that $|x - x_0| \leq 2^{-j_0}$. If x_0 is such a point, using Theorem 2 for all couples $(j, k_j(x_0))$ such that $j \geq j_0$, we have $|C_{j,k_j(x_0)}| \leq C2^{-\gamma j}$, which, using (4.23), can be rewritten

$$\forall j \geq 0, \qquad |2^{-\gamma j}\chi_{j,k_j(x_0)} + f_{j,k_j(x_0)}| \leq C2^{-\gamma j},$$

where $f_{j,k}$ denotes the wavelet coefficients of f. Let

$$p_{j,k} = \mathbb{P}(|C_{j,k}| \leq C2^{-\gamma j}).$$

Then

$$p_{j,k} = \sqrt{\frac{2}{\pi}} \int_{-C-2^{\gamma j}f_{j,k}}^{C-2^{\gamma j}f_{j,k}} e^{-x^2/2}dx \leq 2C\sqrt{\frac{2}{\pi}} := p_C.$$

The end of the proof follows just as in the Brownian case.

5   The multifractal formalism

A common feature shared by (2.4), (2.5), (4.14) and (4.21) is that all the pointwise exponents we considered can be deduced from a countable number of quantities indexed by the dyadic cubes, and the derivation is performed on a log-log plot bearing on the cubes that contain the point x_0. Therefore, these examples all fit in the following general framework for sets of positive quantities e_λ indexed by a subset of the dyadic cubes, for which we introduce the following definition:

- The e_λ are *hierarchical* if there exists $C > 0$ such that, if λ' is a "child" of λ (i.e. $\lambda' \subset \lambda$ and $j' = j+1$) then $e_{\lambda'} \leq C \cdot e_\lambda$.
- The e_λ are *strictly hierarchical* if $\lambda' \subset \lambda \Rightarrow e_{\lambda'} \leq e_\lambda$.
- The (e_λ) belong to $C^\alpha(x_0)$ if, for j large enough, $e_{\lambda_j(x_0)} \leq 2^{-\alpha j}$.
- The (e_λ) belong to $I^\alpha(x_0)$ if, for j large enough, $e_{\lambda_j(x_0)} \geq 2^{-\alpha j}$.
- The *pointwise exponent* associated with the e_λ is

$$h(x_0) = \sup\{\alpha : (e_\lambda) \in C^\alpha(x_0)\} = \liminf_{j \to +\infty} \left(\frac{\log\left(e_{\lambda_j(x_0)}\right)}{\log(2^{-j})} \right). \quad (5.1)$$

- The *upper pointwise exponent* is

$$\tilde{h}(x_0) = \inf\{\alpha : (e_\lambda) \in I^\alpha(x_0)\} = \limsup_{j \to +\infty} \left(\frac{\log\left(e_{\lambda_j(x_0)}\right)}{\log(2^{-j})} \right). \quad (5.2)$$

- The *structure function* is

$$S_j(p) = 2^{-dj} \sum_{\lambda \in \Lambda_j} (e_\lambda)^p.$$

- The scaling function and the upper scaling function are respectively

$$\eta(p) = \liminf_{j \to +\infty} \left(\frac{\log\left(S_j(p)\right)}{\log(2^{-j})} \right) \quad \text{and} \quad \tilde{\eta}(p) = \limsup_{j \to +\infty} \left(\frac{\log\left(S_j(p)\right)}{\log(2^{-j})} \right).$$
(5.3)

- The discrete Besov spaces $\mathcal{B}_p^s$ are defined (for $s, p \in \mathbb{R}$) by

$$(e_\lambda)_{\lambda \in \Lambda} \in \mathcal{B}_p^s \iff \exists C \; \forall j : \quad 2^{-dj} \sum_{\lambda \in \Lambda_j} (e_\lambda)^p \leq C \cdot 2^{-spj}.$$
(5.4)

- The upper Besov spaces $\tilde{\mathcal{B}}_p^s$ are defined (for $s, p \in \mathbb{R}$) by

$$(e_\lambda)_{\lambda \in \Lambda} \in \tilde{\mathcal{B}}_p^s \iff \exists C \; \exists j_n \to +\infty : \; 2^{-dj_n} \sum_{\lambda \in \Lambda_{j_n}} (e_\lambda)^p \leq C \cdot 2^{-spj_n}.$$
(5.5)

Here Λ_j is a subset of the set of dyadic cubes of width 2^{-j}. In practice, it will usually consist in all the dyadic cubes included in a given bounded domain, and therefore this set will be finite, of cardinality $\sim 2^{dj}$.

The irregularity condition $I^\alpha(x_0)$ has been considered since the 70's in the Hölder setting in order to study irregularity properties of stochastic processes, and more recently by M. Clausel in [9]. In the measure setting, it has been considered by Brown, Michon and Peyrière in [6] and by Tricot in [39, 40].

The purpose of introducing strictly hierarchical sequences·is that it is a condition satisfied by sequences of wavelet leaders. It will be particularly relevant in Section 5.5 where we construct counterexamples to the multifractal formalism.

Note that the definition of η that we give here is in agreement with (1.6). Among the above definitions of discrete Besov spaces, the only one which defines a vector space is (5.4) when p is positive. In that case, $\mathcal{B}_p^s$ is a Banach space if $p \geq 1$ and, if $0 < p < 1$, it is a quasi-Banach space endowed with a metric defined as follows: Let $e = (e_\lambda)$ and $f = (f_\lambda)$ be two sequences in $\mathcal{B}_p^s$; then

$$dist(e, f) = \sup_{j \geq 0} \left(2^{(sp-d)j} \sum_k (|e_\lambda - f_\lambda|)^p \right).$$
(5.6)

The spaces $\mathcal{B}_p^s$ are closely related with Besov spaces; indeed, if $q = +\infty$, a function f belongs to $B_p^{s,\infty}$ if $(c_k) \in l^p$ and if its wavelet coefficients c_λ satisfy (5.4). Therefore the wavelet decomposition establishes an isomorphism between the discrete and the continuous Besov spaces.

If the (e_λ) are the wavelet leaders of f, (5.4) yields the wavelet characterization of the oscillation spaces considered in [20].

Note that

$$
\left.\begin{array}{l}
\text{if } p > 0, \ \eta(p) = \sup\{s : \ f \in \mathcal{B}_p^{s/p}\} \ \text{ and } \ \tilde{\eta}(p) = \sup\{s : \ f \in \tilde{\mathcal{B}}_p^{s/p}\}; \\
\text{if } p < 0, \ \eta(p) = \inf\{s : \ f \in \mathcal{B}_p^{s/p}\} \ \text{ and } \ \tilde{\eta}(p) = \inf\{s : \ f \in \tilde{\mathcal{B}}_p^{s/p}\}.
\end{array}\right\}
$$
$$(5.7)$$

In the measure case, $e_\lambda = \mu(3\lambda)$, in the pointwise Hölder case, $e_\lambda = d_\lambda$, and in the $T_\alpha^p(x_0)$ case, $e_\lambda = d_\lambda^p$, so that the general setting in which we work in this section covers all the previous cases we already considered. Therefore, the results we will obtain in this general setting will be valid for all the settings we previously considered: measures, Hölder exponent, p-exponent. However, more precise properties can hold in these particular settings since the corresponding sequences satisfy additional properties (for instance, the $\mu(3\lambda)$ and the d_λ are strictly hierarchical).

We will reformulate the fundamental idea due to G. Parisi and U. Frisch in this general setting. They gave an interpretation of the nonlinearity of the scaling function as the signature of the presence of different pointwise exponents (see [37] and also [14] for applications, particularly in the setting of invariant measures of dynamical systems).

In cases where a whole range of pointwise exponents are present, we will see that the scaling function gives an information about the size of the set of points that display exactly this exponent. A first question is to determine what precise mathematical meaning should be given to the word "size" in this context. Assume for instance that all exponents in the range $[h_{\min}, h_{\max}]$ are obtained, with $h_{\min} < h_{\max}$, and let E_H denote the corresponding isohölder sets, i.e.

$$
E_H = \{x : \ h(x) = H\}.
$$

The first idea is to try the most usual mathematical meaning of "size" in analysis, i.e. the Lebesgue measure; however, this cannot be the right notion here; indeed, because of the countable additivity of the Lebesgue measure, and since bounded domains have a finite measure, it follows that $meas(E_H) = 0$ for almost all values of H. A standard way in order to compare the sizes of different sets that have a vanishing Lebesgue measure is to compute their *fractal dimension*. The idea of associating fractal sets to measures or functions can be traced back to the works of B. Mandelbrot in the 1970s and 1980s [30, 31]. Independently, Orey and Taylor in [35] were the first to consider dimensions of the sets of points where the modulus of continuity of the Brownian motion has a particular order of magnitude (slow and fast points of Brownian motion), which is very close to the idea of multifractal analysis.

5.1 Fractal dimensions and spectrums of singularities

We recall the different notions of dimension which are used. The simplest one is the *box dimensions*.

Definition 18. *Let $A \subset \mathbb{R}^d$; if $\epsilon > 0$, let $N_\epsilon(A)$ be the smallest number of sets of radius ϵ required to cover A.*

The upper box dimension of A is

$$\overline{dim}_B(A) = \limsup_{\epsilon \to 0} \frac{\log N_\epsilon(A)}{-\log \epsilon}.$$

The lower box dimension of A is

$$\underline{dim}_B(A) = \liminf_{\epsilon \to 0} \frac{\log N_\epsilon(A)}{-\log \epsilon}.$$

One important drawback of the box dimensions is that, if A is dense, then the box dimensions take invariably the value d. Since most multifractal functions of interest have dense sets of Hölder singularities, box dimensions are unable to draw any distinction between the sizes of these sets. This explains why box dimensions are not used in the definition of the spectrum of singularities (see Definition 20). However, they are an intermediate step in the definition of the packing dimension that we give below.

Two alternative definitions of fractal dimension have been introduced and are used in multifractal analysis. In order to define the Hausdorff dimensions, we need to recall the notion of δ-dimensional Hausdorff measure.

Definition 19. *Let $A \subset \mathbb{R}^d$. If $\epsilon > 0$ and $\delta \in [0, d]$, we denote*

$$M_\epsilon^\delta = \inf_R \left(\sum_i |A_i|^\delta \right),$$

where R is an ϵ-covering of A, i.e. a covering of A by bounded sets $\{A_i\}_{i \in \mathbb{N}}$ of diameters $|A_i| \leq \epsilon$. The infimum is therefore taken on all ϵ-coverings.

For any $\delta \in [0, d]$, the δ-dimensional Hausdorff measure of A is

$$mes_\delta(A) = \lim_{\epsilon \to 0} M_\epsilon^\delta.$$

There exists $\delta_0 \in [0, d]$ such that

$$\forall \delta < \delta_0, \quad mes_\delta(A) = +\infty$$
$$\forall \delta > \delta_0, \quad mes_\delta(A) = 0.$$

This critical δ_0 is called the Hausdorff dimension of A, and is denoted by $dim(A)$.

The other notion of dimension we will use is the *packing dimension* which was introduced by C. Tricot, see [39, 40] (see also Chap. 5 of [33]). The *lower packing dimension* is

$$Dim(A) = \inf \left\{ \sup_{i \in \mathbb{N}} \left(\underline{dim}_B A_i : A \subset \bigcup_{i=1}^{\infty} A_i \right) \right\} \qquad (5.8)$$

(the infimum is taken over all possible partitions of A into a countable collection A_i). We will use this alternative notion in order to bound the dimensions of some sets of singularities. The dimensions we introduced can be compared as follows, see [33, 39, 40],

$$\forall A \subset \mathbb{R}^d, \qquad dim(A) \leq Dim(A) \leq \underline{dim}_B(A) \leq \overline{dim}_B(A). \qquad (5.9)$$

A usual way to classify fractal sets in mathematics is to consider their Hausdorff dimensions; this motivates the following definition.

Definition 20. *Let* $(e_\lambda)_{\lambda \in \Lambda}$ *be a hierarchical dyadic function, and let*

$$E_H = \{x : \quad h(x) = H\}.$$

The spectrum of singularities associated with the $(e_\lambda)_{\lambda \in \Lambda}$ *is the function* d *defined by*

$$d(H) = dim(E_H)$$

(we use the convention $d_f(H) = -\infty$ *if* H *is not a Hölder exponent of* f*). The support of the spectrum is the set of values of* H *for which* $E_H \neq \emptyset$.

Remark: We can consider many alternative definitions which are variants of this one and are obtained by considering the set

$$F_H = \{x : \quad \tilde{h}(x) = H\}.$$

We can also consider the sets of points where these exponents are larger or smaller than a given value, and finally the corresponding Hausdorff or packing dimensions. This leads to a large collection of possible spectra. They will be useful in Section 5.4 in order to obtain optimal upper bounds for the dimensions of the spectra.

The spectrum of singularities of many mathematical functions or measures can be determined directly from its definition. On the opposite, for many real-life signals, whose Hölder exponent is expected to be everywhere discontinuous, the numerical determination of their Hölder regularity is not feasible, and therefore, one cannot expect to have direct access to their spectrum of singularities. In such cases, one has to find an indirect way to compute $d(H)$; the multifractal formalism is a formula which is expected to yield the spectrum of singularities of f from the scaling function, which is numerically computable (We will derive such a formula in Section 5.2). Mathematically, these quantities are interpreted as indicating that f belongs to a certain subset of a family of function spaces.

The pointwise exponent of a sequence often takes all values on a whole interval $[h_{\min}, h_{\max}]$, where $h_{\min} < h_{\max}$; in such cases, the computation of the spectrum of singularities requires the study of an infinite number of fractal sets E_H; this explains the introduction of the term "multifractal" by G. Parisi and U. Frisch in [37]. Note that speaking of the "multifractal analysis" of either a mathematical function or a signal derived from real-life data does not imply that it is assumed to be multifractal: Such an analysis often concludes that it is monohölder for instance.

5.2   Derivation of the multifractal formalism

Let us now show how the spectrum of singularities is expected to be recovered from the scaling function. The definition of the scaling function (5.3) roughly means that, for j large,

$$S_j(p) \sim 2^{-\eta(p)j}.$$

Let us estimate the contribution to $S_j(p)$ of the dyadic cubes λ that cover the points of E_H. By definition of E_H, they satisfy

$$e_\lambda \sim 2^{-Hj};$$

by definition of $d(H)$, since we use cubes of the same width 2^{-j} to cover E_H, we need about $2^{-d(H)j}$ such cubes; therefore the corresponding contribution is of the order of magnitude of

$$2^{-dj}2^{d(H)j}2^{-Hpj} = 2^{-(d-d(H)+Hp)j}.$$

When $j \to +\infty$, the dominant contribution comes from the smallest exponent, so that

$$\eta(p) = \inf_H(d - d(H) + Hp). \tag{5.10}$$

We will show that the scaling function $\eta(p)$ is a concave function on $\mathbb{R}$, which is in agreement with the fact that the right-hand side of (5.10) necessarily is a concave function (as an infimum of a family of linear functions) no matter whether $d(H)$ is concave or not. However, if $d(H)$ also is a concave function, then the Legendre transform in (5.10) can be inverted (as a consequence of the duality of convex functions), which justifies the following assertion:

Definition 21. *A sequence* (e_λ) *follows the multifractal formalism if its spectrum of singularities satisfies*

$$d(H) = \inf_{p \in \mathbb{R}} (d - \eta(p) + Hp). \tag{5.11}$$

Note that an alternative method in order to recover the spectrum is proposed in [8]; it is commonly used in practice since it allows to get rid of the Legendre transform.

The derivation exposed above is not a mathematical proof, and the determination of the range of validity of (5.11) (and of its variants) is one of the main mathematical problems concerning multifractal analysis. It does not hold in complete generality. However, three types of verifications can be performed:

- The multifractal formalism is proved under additional assumptions on the e_λ (usually, assuming that it is derived from a self-similar function or measure).

- It is proved for a "large" subset of the function space considered: We refer to [13, 20] and references therein for "generic results" of multifractality, either in the sense either of Baire categories or of prevalence.

- The multifractal formalism is shown to yield an upper bound of the spectrum of singularities, see Section 5.4.

Note that, in applications, it often happens that the spectrum of singularities itself has no direct scientific interpretation and multifractal analysis is only used as a classification tool in order to discriminate between several types of signals; then, one is no more concerned with the validity of (5.11) but only with having its right-hand side defined in a numerically precise way.

5.3 Properties of the scaling function

We start by proving that the function η is concave on $\mathbb{R}$, a property that was used in the derivation of the multifractal formalism

Proposition 4. *The function η defined by (5.3) is concave on $\mathbb{R}$.*

Note that $\tilde{\eta}$ (also defined by (5.3)) has no reason to be concave in general. However, in practical applications, the liminf and limsup in (5.3) often coincide (and are therefore true limits), in which case $\tilde{\eta} = \eta$ and is therefore concave. In order to prove Proposition 4, we will need the following lemma.

Lemma 5. *Let $(a_i)_{i \in \mathbb{N}}$ be a sequence of positive real numbers. The function $\omega : \mathbb{R} \longrightarrow \bar{\mathbb{R}} \ (= \mathbb{R} \cup \{+\infty, -\infty\})$ defined by*

$$\omega(p) = \log\left(\sum_{i \in \mathbb{N}} a_i^p\right)$$

is convex on $\mathbb{R}$.

Proof of Lemma 5: We need to check that

$$\forall p, q \in \mathbb{R}, \quad \forall \alpha \in]0, 1[, \quad \omega\left(\alpha p + (1-\alpha)q\right) \leq \alpha\omega(p) + (1-\alpha)\omega(q). \tag{5.12}$$

Consider the sequences

$$A = (a_1^{\alpha p}, ...a_N^{\alpha p}, \cdots) \quad \text{and} \quad B = (a_1^{(1-\alpha)q}, ...a_N^{(1-\alpha)q}, \cdots),$$

Hölder's inequality applied with the conjugate exponents $p' = 1/\alpha$ and $q' = 1/(1 - \alpha)$ yields

$$\sum_{i=1}^{\infty} a_i^{\alpha p + (1-\alpha)q} \leq \left(\sum_{i=1}^{\infty} a_i^p\right)^{\alpha} \left(\sum_{i=1}^{\infty} a_i^q\right)^{1-\alpha}.$$

Taking logarithms on both sides of this inequality yields (5.12).

We will now show that η is concave on $\mathbb{R}$. For each j, one applies Lemma 5 to the e_λ. We obtain that, $\forall j$, the function

$$p \to \log\left(\sum_{\lambda \in \Lambda_j} d_\lambda^p\right)$$

is convex: therefore, after dividing by $\log(2^{-j})$, we obtain a concave function: since concavity is preserved by taking infimums and pointwise limits, the concavity of the scaling function follows.

We will make the following uniform regularity and irregularity assumptions:

$$\exists C_1, C_2, A, B, \quad \text{such that} \quad \forall \lambda \quad C_1 2^{-Bj} \leq e_\lambda \leq C_2 2^{-Aj}. \tag{5.13}$$

Note that in the measure case and in Hölder exponent case, one can pick $A = 0$. In the Hölder case, the uniform regularity assumption means that $A > 0$. When the e_λ are wavelet leaders, the assumption on the lower bound implies that the function f considered has no C^∞ components. In the measure case, it implies that μ does not vanish on a set of nonempty interior.

In order to better understand the range of validity of the multifractal formalism, we have to explore the properties of the Legendre transform of the scaling function. Therefore, we introduce the *Legendre spectrum* associated with the sequence (e_λ):

$$L(H) = \inf_{p \in \mathbb{R}} (d + Hp - \eta(p)). \tag{5.14}$$

The validity of the multifractal formalism therefore states that the Legendre spectrum coincides with the spectrum of singularities.

Lemma 6. *Let (e_λ) be a sequence satisfying (5.13);*

$$\text{if} \quad p \geq 0, \quad \text{then} \quad Ap \leq \eta(p) \leq Bp,$$

$$\text{if} \quad p \leq 0, \quad \text{then} \quad Bp \leq \eta(p) \leq Ap.$$

Proof of Lemma 6: The assumption (5.13) implies that, if $p > 0$,

$$(C_1)^p 2^{-Bpj} \leq S_j(p) \leq (C_2)^p 2^{-Apj},$$

and the inequalities are reversed if $p < 0$. The result follows from the definition of the scaling function (5.3).

It follows from Lemma 6 that $\eta(0) = 0$, $L(H) = -\infty$ if $H \leq A$ or $H \geq B$, and

$$\forall p \in \mathbb{R}, \quad A \leq \eta'(p) \leq B.$$

Proposition 5. *Let (e_λ) be a sequence satisfying (5.13). Let*

$$H_{\min} = \sup\{A : \ (5.13) \ holds\} \quad and \quad H_{\max} = \inf\{B : \ (5.13) \ holds\}.$$

The Legendre spectrum $L(H)$ of (e_λ) is a concave function satisfying

$$\text{if} \quad H \notin [H_{\min}, H_{\max}], \quad L(H) = -\infty, \tag{5.15}$$

$$\text{if} \quad H \in [H_{\min}, H_{\max}], \quad 0 \leq L(H) \leq d. \tag{5.16}$$

There exist H^1_{med} and $H^2_{\text{med}} \in [H_{\min}, H_{\max}]$ such that

- *$L(H)$ is strictly increasing on $[H_{\min}, H^1_{\text{med}}]$,*
- *$\forall H \in [H^1_{\text{med}}, H^2_{\text{med}}]$, $L(H) = d$,*
- *$L(H)$ is strictly decreasing on $[H^2_{\text{med}}, H_{\max}]$.*

Remark: It follows immediately from the Legendre transform formula (5.14) that $H^1_{\text{med}} = \eta'_r(0)$ and $H^2_{\text{med}} = \eta'_l(0)$ (where η'_r and η'_l respectively denote the right and left derivatives of η); therefore, if η is differentiable at 0, then H^1_{med} and H^2_{med} coincide; in that case, we will denote their common value by H_{med}. Formula (5.16) means that the multifractal formalism cannot yield "negative" dimensions, when applied to one given set of data. Such a phenomenon can only happen after an averaging over a large number of realizations, in a probabilistic setting, i.e. when defining the scaling function as $\mathbb{E}((e_\lambda)^p)$ and not sample path by sample path.

We will call *admissible spectral function* any function $L(H)$ which satisfies the conditions listed in Proposition 5. An *admissible scaling function* is any function $\eta(p)$ which is the Legendre transform of an admissible spectral function. Theorem 6 will imply that these conditions characterize the functions $L(H)$ which are the Legendre spectra of a given sequence (e_λ); and, equivalently, they also characterize the functions $\eta(p)$ which are scaling functions.

Proof of Proposition 5: The fact that $L(H)$ is concave follows directly from its definition as a Legendre transform. Lemma 6 implies that, if $p \geq 0$, then $\eta(p) \geq H_{\mathrm{min}}p$ and, if $p \leq 0$, then $\eta(p) \geq H_{\mathrm{max}}p$; (5.15) follows from these lower bounds.

Since $\eta(0) = 0$, it follows that $\forall H$, $L(H) \leq d$, and, the fact that $\forall H \in [H_{\mathrm{med}}^1, H_{\mathrm{med}}^2]$, $L(H) = d$ directly follows from (5.14). The increasing and then decreasing property follows from this observation together with the concavity property.

Let us now prove that $L(H)$ is nonnegative on $[H_{\mathrm{min}}, H_{\mathrm{max}}]$. From the definition of $L(H)$, it suffices to prove that

$$\forall H \in (H_{\mathrm{min}}, H_{\mathrm{max}}), \ \forall p \in \mathbb{R}, \qquad\qquad \eta(p) \leq d + Hp \qquad (5.17)$$

(indeed, by continuity of L, the result will also be true for the extreme points H_{min} and H_{max}, if L differs from $-\infty$ at these points). By definition of H_{min} and H_{max}, there exist sequences $l_n, j_n \in \mathbb{N}$, and dyadic cubes μ_n and λ_n such that

$$|\mu_n| = 2^{-l_n}, \quad |\lambda_n| = 2^{-j_n}, \quad e_{\mu_n} \leq 2^{-Hl_n}, \quad \text{and} \quad e_{\lambda_n} \geq 2^{-Hj_n}.$$

Therefore, if $p < 0$,

$$2^{-dl_n} \sum_{\lambda \in \Lambda_{l_n}} |e_\lambda|^p \geq 2^{-(d+Hp)l_n}$$

and if $p > 0$,

$$2^{-dj_n} \sum_{\lambda \in \Lambda_{j_n}} |e_\lambda|^p \geq 2^{-(d+Hp)j_n},$$

and (5.17) follows.

5.4   Upper bound of the spectrums

Corollary 2 will state that $L(H)$ yields upper bound for the spectrum of singularities of any sequence. It will be a straightforward consequence of the following sharper bounds that involve the different types of singularities we introduced.

Theorem 5. *Let* (e_λ) *be a sequence and*

$$J^H = \{x_0 : (e_\lambda) \in C^H(x_0)\}, \qquad G^H = (J^H)^c,$$
$$F^H = \{x_0 : (e_\lambda) \in I^H(x_0)\}, \qquad K^H = (F^H)^c.$$

- *If* $(e_\lambda)_{\lambda \in \Lambda} \in \mathcal{B}_p^s$ *with* $p > 0$, *then* $dim(G^H) \leq d - sp + Hp$.
- *If* $(e_\lambda)_{\lambda \in \Lambda} \in \tilde{\mathcal{B}}_p^s$ *with* $p > 0$, *then* $Dim(F^H) \leq d - sp + Hp$.
- *If* $(e_\lambda)_{\lambda \in \Lambda} \in \mathcal{B}_p^s$ *with* $p < 0$, *then* $dim(K^H) \leq d - sp + Hp$.
- *If* $(e_\lambda)_{\lambda \in \Lambda} \in \tilde{\mathcal{B}}_p^s$ *with* $p < 0$, *then* $Dim(J^H) \leq d - sp + Hp$.

Remark: Using (2.4) and (2.5), this theorem, if applied to the sequence $e_\lambda = \mu(3\lambda)$, allows to recover the bounds for either packing or Hausdorff dimensions of the sets of points where the exponents h_μ and $\overline{h}_\mu$ of a measure μ are larger or smaller than H, see [6]. Similarly, if f is a uniform Hölder function, (4.14) shows that Theorem 5, if applied to the sequence $e_\lambda = d_\lambda$ yields the same bounds for the Hölder exponent. Finally, if f belongs to L^p, Theorem 3 implies a similar result for the p-exponent, using the sequence $e_\lambda = d_\lambda^p$.

In order to prove this theorem, we start by establishing simple estimates on the number of small and large coefficients e_λ.

Lemma 7. *Let*

$$\begin{cases} G_{j,H} = \{\lambda : e_\lambda \geq 2^{-Hj}\}, \, N_{j,H} = Card(G_{j,H}) \\ F_{j,H} = \{\lambda : e_\lambda \leq 2^{-Hj}\}, \, M_{j,H} = Card(F_{j,H}). \end{cases}$$

- *If* $(e_\lambda)_{\lambda \in \Lambda} \in \mathcal{B}_p^s$ *with* $p > 0$, *then* $\forall j$, $N_{j,H} \leq 2^{(d-sp+Hp)j}$.
- *If* $(e_\lambda)_{\lambda \in \Lambda} \in \tilde{\mathcal{B}}_p^s$ *with* $p > 0$, *then* $\exists j_n \to \infty$: $N_{j_n,H} \leq 2^{(d-sp+Hp)j_n}$.
- *If* $(e_\lambda)_{\lambda \in \Lambda} \in \mathcal{B}_p^s$ *with* $p < 0$, *then* $\forall j$, $M_{j,H} \leq 2^{(d-sp+Hp)j}$.
- *If* $(e_\lambda)_{\lambda \in \Lambda} \in \tilde{\mathcal{B}}_p^s$ *with* $p < 0$, *then* $\exists j_n \to \infty$: $M_{j_n,H} \leq 2^{(d-sp+Hp)j_n}$.

Proof of Lemma 7: Let us consider the first assertion. Since, $(e_\lambda)_{\lambda \in \Lambda} \in \mathcal{B}_p^s$ with $p > 0$,

$$\exists C, \, \forall j, \quad 2^{(sp-d)j} \sum |e_\lambda|^p \leq C;$$

therefore, by restricting the sum to the elements of $G_{j,H}$, we obtain

$$\exists C,\ \forall j, \quad 2^{(sp-d)j} N_{j,H} 2^{-Hpj} \leq C,$$

so that $N_{j,H} \leq C 2^{(d-sp+Hp)j}$. The proof of the third case is similar because, if p is negative, the condition $e_\lambda \leq 2^{-Hj}$ becomes $(e_\lambda)^p \geq 2^{-Hpj}$.

The second and the fourth case also follow by the same arguments applied to a subsequence j_n.

Let (A_j) be a sequence of sets; we recall that $\overline{\lim}(A_j)$ denotes the set of points that belong to an infinite number of the A_j and $\underline{\lim}(A_j)$ denotes the set of points that belong to all A_j for j large enough. It follows from the definitions of G_H and K_H that

$$G_H = \overline{\lim}(G_{j,H}) \quad \text{and} \quad K_H = \overline{\lim}(F_{j,H}).$$

Furthermore, if j_n is the sequence that appears in the definition of $\tilde{B}_p^s$, then it follows from the definitions of J_H and F_H that

$$F_H = \underline{\lim}(G_{j,H}) \subset \underline{\lim}(G_{j_n,H}) \quad \text{and} \quad J_H = \underline{\lim}(F_{j,H}) \subset \underline{\lim}(F_{j_n,H}).$$

Therefore, in order to prove Theorem 5, there only remains to prove the following estimates on Hausdorff and Packing dimensions.

Lemma 8. *Let $j_n \to +\infty$ and A_{j_n} be a union of at most $2^{\omega j_n}$ dyadic cubes $\lambda_{j_n,k}$ of width 2^{-j_n}. Then*

$$dim\left(\overline{\lim}(A_{j_n})\right) \leq \omega \quad \text{and} \quad Dim\left(\underline{\lim}(A_{j_n})\right) \leq \omega.$$

Proof: We use $\bigcup_{l \geq n} A_{j_l}$ for covering of $\overline{\lim}(A_{j_n})$; it is an ϵ-covering as soon as $\sqrt{d} 2^{-j_n} \leq \epsilon$. It follows that, for this covering,

$$\sum_{l \geq n} \sum_k (diam(\lambda_{j_l,k}))^\delta \leq \sum_{l \geq n} 2^{\omega j_l} 2^{-\delta j_l}$$

which is finite as soon as $\delta > \omega$.

Note that $\underline{\lim}(A_{j_n})$ is the countable union of the sets

$$B_{j_n} = \bigcap_{l \geq n} A_{j_l}.$$

For any $l \geq n$, B_{j_n} is included in A_{j_l}, i.e. in the union of $2^{\omega j_l}$ dyadic cubes of width 2^{-j_l}, so that the lower box dimension of B_{j_n} is bounded by ω, and the lower packing dimension of $\underline{\lim}(A_{j_n})$ is bounded by ω.

Corollary 2. *Let (e_λ) be a sequence. Then*

$$d(H) \leq \inf_{p \in \mathbb{R}} \left(Hp - \eta(p) + d \right). \qquad (5.18)$$

Proof: Clearly, $\forall \epsilon > 0$, $E_h \subset G^{H+\epsilon}$. Furthermore, using (5.7),

$$\forall p > 0, \ \forall \delta > 0, \quad (e_\lambda) \in B_p^{\frac{\eta(p)}{p} - \delta},$$

so that, by Theorem 5,

$$dim(E_H) \leq d - \left(\frac{\eta(p)}{p} - \delta \right) p + (H + \epsilon)p.$$

Since this is true $\forall \epsilon, \delta > 0$, we obtain the expected bound for $p > 0$. Similarly, we have $\forall \epsilon > 0$, $E_h \subset J^{H-\epsilon}$, and using (5.7),

$$\forall p < 0, \ \forall \delta > 0, \quad (e_\lambda) \in B_p^{\frac{\eta(p)}{p} + \delta},$$

so that, by Theorem 5,

$$dim(E_H) \leq d - \left(\frac{\eta(p)}{p} + \delta \right) p + (H - \epsilon)p.$$

Since this is true $\forall \epsilon, \delta > 0$, we obtain that

$$\forall p < 0, Dim(E_H) \leq Hp - \eta(p) + d;$$

the required bound for $p < 0$ follows from (5.9).

5.5   Validity of the multifractal formalism

The following theorem states that, except for the very particular case of a linear scaling function, the multifractal formalism is not valid unless additional properties are assumed for the e_λ.

Theorem 6. *Let (e_λ) be a sequence. If their scaling function η is linear, i.e. if*

$$\exists \alpha > 0 \quad \text{such that} \quad \forall p \in \mathbb{R}, \quad \eta(p) = \alpha p, \qquad (5.19)$$

then the multifractal formalism holds for the (e_λ), and their spectrum of singularities is given by

$$\begin{cases} d(\alpha) = 1 \\ d(H) = -\infty \quad \text{if} \quad H \neq \alpha. \end{cases}$$

If η is a nonlinear scaling function, then there exists a sequence (e_λ)
whose scaling function is η, and whose spectrum of singularities is given
by:

$$\begin{cases} d(H_{\min}) = d(H_{\max}) = 0 \\ d(H_{\mathrm{med}}^1) = d(H_{\mathrm{med}}^2) = 1 \\ d(H) = -\infty \quad \textit{if} \quad H \notin \{H_{\min}, H_{\mathrm{med}}^1, H_{\mathrm{med}}^2, H_{\max}\}, \end{cases} \tag{5.20}$$

where $H_{\min}, H_{\mathrm{med}}^1, H_{\mathrm{med}}^2,$ and $H_{\max}$ are defined in Proposition 5.

Remark: A sequence whose spectrum is given by (5.20) clearly is
a counterexample to the multifractal formalism. Note that, though, the
linear case given by (5.19) can seem exceptional, it covers important examples such as the Weierstrass functions, or the FBM, see [1]. We state
and prove Theorem 6 in dimension one; one easily deduce counterexamples in several dimensions.

Figure 5.1 shows the function whose wavelet coefficients (and wavelet
leaders) are the (e_λ) constructed in Proof of Theorem 6 for the counterexample with $H_{\min} = 0.6$, $H_{\mathrm{med}}^1 = H_{\mathrm{med}}^2 = 0.8$ and $H_{\max} = 1$. This
function has the same Legendre spectrum that an MRW — for which the
multifractal formalism is valid - but not the same spectrum of singularity.

Proof of Theorem 6: Assume that the scaling function is linear
and given by

$$\eta(p) = \alpha p.$$

Then $L(H) = -\infty$ except for $H = \alpha$; Corollary 2 implies in this case
that $d(H) \leq -\infty$ for $H \neq \alpha$; therefore only one Hölder exponent is
present, so that $\forall x,\ h(x) = \alpha$; it follows that $d(\alpha) = 1$, and the multifractal formalism therefore holds.

In order to construct a counterexample in the nonlinear case, we will
define explicitly the e_λ at each scale for $\lambda \subset [0,1]$. We fix a $j \geq 0$.
The $e_\lambda = e_{j,k}$ are defined from $k = 0$ to $2^j - 1$ and are increasing when
k increases. For $k = 0$ to $\lceil 2^{L(H_{\max})j} \rceil$ we define $e_{j,k} = 2^{-H_{\max}j}$. When
H decreases form $H_{\max}$ to H_{med}^2 the function $2^{L(H)j}$ is continuous and
increases up to 2^j. Each time it reaches a new integer value k we define
H_k as the only $H \in (H_{\mathrm{med}}^2, H_{\max}]$ such that

$$2^{L(H_k)j} = k;$$

the value of the k-th coefficient is

$$e_{j,k} = 2^{-H_k j}.$$

We do this until we have reached the value $k = 2^{j-1} - 1$. After this, we
jump to the value of $H \in [H_{\min}, H_{\mathrm{med}}^1)$ such that $L(H) = 2^{j-1}$. We con-
tinue similarly: When H decreases, each time the function $2^{L(H)j}$ reaches

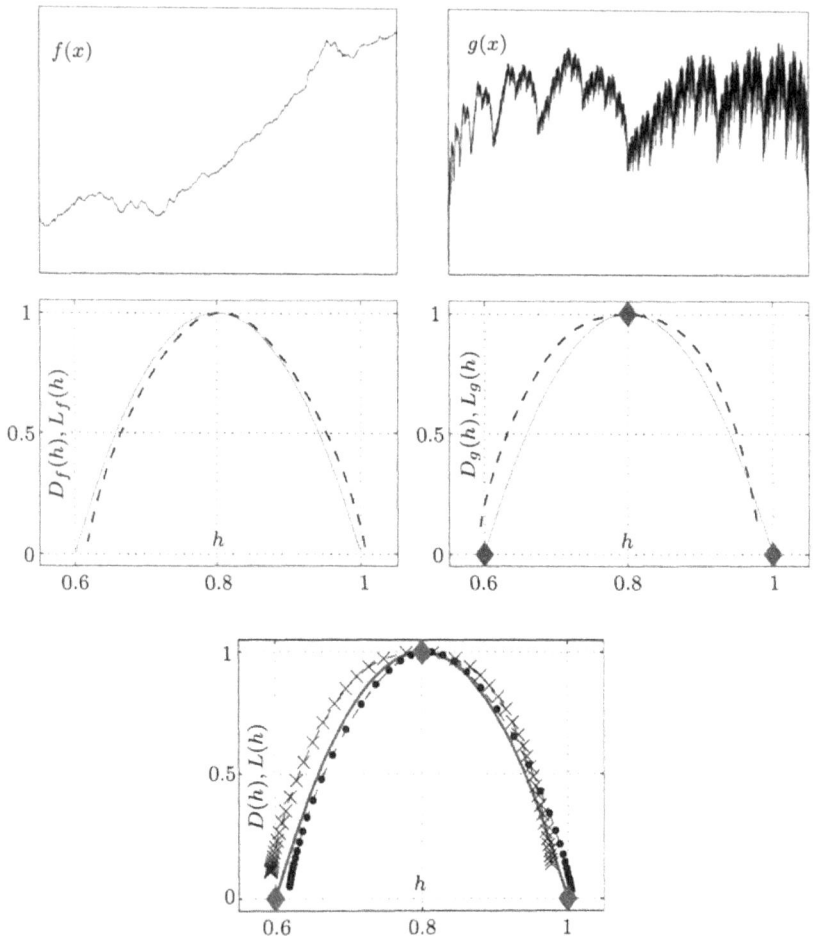

Figure 5.1   **Validity of the multifractal formalism.** Sample of MRW (top left) and process contructed according to the counter example for Theorem 6 (top right). Both functions have identical Legendre spectra but different singularity spectra. Middle left: Singularity and Legendre spectra (red solid line) are equal for MRW, their leader based estimate is superimposed (black dashed line with dots). Middle right: Singularity spectrum (red diamonds) and Legendre spectrum (red solid line) together with the leader based estimate (black dashed line with crosses). Bottom: All spectra are superimposed: Leader based estimates for MRW an for the counter example are very close and satisfactorily match the common Legendre spectrum of both processes. While for MRW, the leader based multifractal formalism enables to estimate the singularity spectrum, this is no longer the case for the counter example tailored in purpose. (*The sample size is $N = 2^{18}$. The counter example is synthesized using Daubechies4 wavelets. Estimations are performed with Daubechies3 wavelets and ordinary linear regressions over scales $j = [2, 10]$. Results for MRW ($c_1 = 0.8, c_2 = -0.02$) are obtained as means over 50 realizations.*)

a new integer value k we define H_{2^j-k} as the only $H \in [H_{\min}, H_{\text{med}}^1]$ such that

$$2^{L(H_{2^j-k})j} = k.$$

The value of the corresponding coefficient is

$$e_{j,2^j-k} = 2^{-H_{2^j-k}j}.$$

When we reach the value $H_{\min}$ we set the remaining coefficients at the value $2^{-H_{\min}j}$.

First, note that all coefficients lie between $2^{-H_{\max}j}$ and $2^{-H_{\min}j}$, so that (5.13) is satisfied. Let us now determine the Hölder exponent of this sequence. By construction, at each scale, there is (at least) one coefficient of size $2^{-H_{\max}j}$ located above $x_0 = 0$ and another coefficient of size $2^{-H_{\min}j}$ located above $x_0 = 1$; therefore, the Hölder exponent at the two extreme points takes the values

$$h(0) = H_{\max} \quad \text{and} \quad h(1) = H_{\min}.$$

For any $\epsilon > 0$, the $2^{L(H_{\text{med}}^2+\epsilon)j}$ first coefficients are smaller than $2^{-(H_{\text{med}}^2+\epsilon)j}$ and the last $2^{L(H_{\text{med}}^1-\epsilon)j}$ coefficients are larger than $2^{-(H_{\text{med}}^1+\epsilon)j}$. As regards the others, either $k \leq 2^{j-1}$ and they lie between $2^{-(H_{\text{med}}^2+\epsilon)j}$ and $2^{-H_{\text{med}}^2 j}$, or $k > 2^{j-1}$ and they lie between $2^{-(H_{\text{med}}^1-\epsilon)j}$ and $2^{-H_{\text{med}}^1 j}$. The length of both intervals tends to $1/2$ because $\forall H < H_{\text{med}}^1$, $L(H) < 1$ and $\forall H > H_{\text{med}}^2$, $L(H) < 1$. Therefore

$$\forall x \in (0, 1/2), \qquad h(x) = H_{\text{med}}^2,$$

$$\forall x \in (1/2, 1), \qquad h(x) = H_{\text{med}}^1,$$

and (5.20) is proved.

Let us now show why the Legendre spectrum of the (e_λ) is indeed $L(H)$. First, we recall the definition of the *large deviation spectrum* $\mathcal{L}(H)$ of a sequence. Let

$$N_j(\epsilon, \alpha) = Card\left(\{\lambda \in \Lambda_j : 2^{-(\alpha+\epsilon)j} \leq |e_\lambda| \leq 2^{-(\alpha-\epsilon)j}\}\right);$$

then

$$\mathcal{L}(H) = \lim_{\epsilon \to 0} \limsup_{j \to +\infty} \frac{\log(N_j(\epsilon, \alpha))}{\log(2^j)}.$$

By construction,

$$N_j(\epsilon, \alpha) = \sup\left(2^{L(\alpha+\epsilon)j}, 2^{L(\alpha-\epsilon)j}\right) + O(1),$$

and therefore a standard large deviation argument yields that there are $\sim 2^{L(h)j}$ coefficients of size $\sim 2^{-Hj}$, and therefore the Legendre spectrum of this sequence is indeed $L(H)$.

Let us now check that the sequence (e_λ) is strictly hierarchical. Let $H > H^2_{\text{med}}$. If $2^{L(H)j} \in \mathbb{N}$, then the coefficient $e_\lambda = e_{j,k}$ of size 2^{-Hj} is such that $k = 2^{L(H)j}$; therefore its distance to the origin is $\delta_\lambda = 2^{-j}2^{L(H)j}$. Since, for each j, the sequence $k \rightarrow e_{j,k}$ is increasing, it suffices to prove that, for the same size of coefficient (i.e. if $Hj = H'j'$), and $j' > j$, then $\delta_{\lambda'} \leq \delta_\lambda$, which is equivalent to

$$2^{(1-L(H'))j'} \geq 2^{(1-L(H))j}.$$

This is, in turn equivalent to prove that, if $H > H'$ then,

$$\frac{1 - L(H')}{H'} \leq \frac{1 - L(H)}{H}$$

which follows from the concavity of $L(H)$, and the fact that it reaches its maximum at H_{med}. We don't treat the case $H < H^1_{\text{med}}$ since it is easier to deal with: indeed it is a straightforward consequence of the fact that $L(H)$ is increasing on $[H_{\text{min}}, H^1_{\text{med}}]$. This last assertion implies that the counterexample we constructed also works for the Hölder exponent of functions. Indeed, if we pick the e_λ for wavelet coefficients of the function, the the wavelet leaders will also be the e_λ.

5.6   Some open questions

One can meet several types of multifractal functions. A first one is supplied by functions which present inhomogeneities: They are smoother in some regions than in others. This case is often met in image analysis. Indeed, a natural image is a patchwork of textures with different characteristics. Their spectrums of singularities reflects the multifractal nature of each component, and also of the boundaries (which may also be fractal) where discontinuities appear. In such situations, the determination of a local spectrum of singularities for each 'component' is more relevant. It is defined as follows: If $\Omega \subset \mathbb{R}^d$ is a nonempty open set,

$$d^\Omega(H) = dim(E_H \cap \Omega);$$

clearly, $d(H) = \sup_\Omega d^\Omega(H)$.

By contrast, *homogeneous* multifractal sequences present the same characteristics everywhere. The same remark applies to the function spaces to which the sequence belongs.

The scaling function $\eta^\Omega(p)$ of a sequence (e_λ) restricted to Ω is defined using (5.3), but starting with the structure function

$$S_j^\Omega(p) = \sum_{\lambda \in \Lambda_j, \, \lambda \subset \Omega} (e_\lambda)^p.$$

For function spaces, one may require a smooth cutoff, so that the scaling function of f restricted to Ω is

$$\eta^{\Omega}(p) = \inf_{\varphi \in \mathcal{D}, supp(\varphi) \subset \Omega} \eta^{\Omega}_{f,\varphi}(p).$$

Definition 22. *A sequence is Hölder-homogeneous if the function $d^{\Omega}(H)$ is independent of Ω. It is scaling-homogeneous if the function $\eta^{\Omega}(p)$ is independent of Ω.*

Note that the same definition holds for functions and measures. The notion of homogeneous function is important for modelling; indeed, with regards to fully developed turbulence, Kolmogorov assumed that the scaling function is *universal*, i.e., it does not depend on the particular fluid considered, on the limit conditions, on the particular region of the fluid considered. Parisi and Frisch also assumed the same property for the spectrum of singularities. Therefore, the velocity of a turbulent fluid is expected to be an homogeneous multifractal function. Usually, the mathematical functions and stochastic processes which are known to be multifractal are actually homogeneous multifractals. An important open question is to determine which functions $d(H)$ can be the spectrum of singularities of an homogeneous multifractal function. Note also that the counterexamples to the multifractal formalism that we gave are not homogeneous. Therefore, another open question is to determine if one can construct counterexamples that are also homogeneous with the same level of generality.

In real-life data obtained one usually observes two properties: The signal is scaling homogeneous, and the scaling function η is actually given in (5.3) by a true limit, which means that $\forall p$, $\eta(p) = \tilde{\eta}(p)$. We do not know what these properties imply on the sequence (e_λ), and in particular, if these additional properties improve the range of validity of the multifractal formalism for such sequences.

References

[1] P. Abry, S. Jaffard and B. Lashermes, *Wavelet leaders in multifractal analysis*, "Wavelet Analysis and Applications", T. Qian et al. eds., pp. 201–246, "Applied and Numerical Harmonic Analysis", Springer (2006).

[2] P. Abry, B. Lashermes and S. Jaffard, *Revisiting scaling, multifractal and multiplicative cascades with the wavelet leader lens*, Optic East, Wavelet applications in Industrial applications II Vol. 5607, pp. 103–117, Philadelphia, USA (2004).

[3] P. Abry, B. Lashermes and S. Jaffard, *Wavelet leader based multifractal analysis*, Proceedings of the 2005 IEEE International Conference on Acoustics, Speech, and Signal Processing.

[4] A. Arneodo, B. Audit, N. Decoster, J.-F. Muzy and C. Vaillant, *Wavelet-based multifractal formalism: applications to DNA sequences, satellite images of the cloud structure and stock market data,* in: The Science of Disasters; A. Bunde, J. Kropp, H. J. Schellnhuber Eds., Springer, pp. 27–102 (2002).

[5] E. Bacry, J. Delour and J. F. Muzy *Multifractal random walk,* Phys. Rev. E, vol. 64, 026103–026106 (2001).

[6] G. Brown, G. Michon and J. Peyrière, *On the multifractal analysis of measures,* J. Statist. Phys., vol. 66, pp. 775–790 (1992).

[7] A. P. Calderòn and A. Zygmund, *Singular integral operators and differential equations,* Amer. J. Math. vol. 79, pp. 901–921 (1957).

[8] A. B. Chhabra, C. Meneveau, R. V. Jensen and K. Sreenivasan, *Direct determination of the $f(\alpha)$ singularity spectrum and its applications to fully developed turbulence,* Phys. Rev. A, vol. 40, pp. 5284–5294, (1989).

[9] M. Clausel, *Etude de quelques notions d'irrégularité: le point de vue ondelettes,* Ph.D. Thesis, Université Paris 12 (2008).

[10] A. Cohen, I. Daubechies and J.-C. Fauveau, *Biorthogonal bases of compactly supported wavelets,* Comm. Pure Appl. Math., vol. 44, pp. 485–560 (1992).

[11] A. Cohen and R. Ryan, *Wavelets and Multiscale Signal Processing,* Chapman and Hall (1995).

[12] J. P. R. Christensen, *On sets of Haar measure zero in Abelian Polish groups,* Israel J. Math. **13** (1972), 255–260.

[13] A. Fraysse and S. Jaffard, *How smooth is almost every function in a Sobolev space?* Rev. Matem. Iberoamer. Vol. 22, N. 2, pp. 663–682 (2006).

[14] T. Halsey, M. Jensen, L. Kadanoff, I. Procaccia and B. Shraiman, *Fractal measures and their singularities: The characterization of strange sets,* Phys. Rev. A, vol. 33, pp. 1141–1151 (1986).

[15] Y. Heurteaux and S. Jaffard, *Multifractal analysis of images: New connexions between analysis and geometry,* Proceedings of the NATO-ASI Conference on Imaging for Detection and Identification, Springer (2007).

[16] B. Hunt, *The prevalence of continuous nowhere differentiable functions,* Proceed. A.M.S **122** (1994), no. 3, 711–717.

[17] B. Hunt and T. Sauer and J. Yorke, *Prevalence: A translation invariant "almost ever" on infinite dimensional spaces,* Bull. A.M.S **27** (1992), 217–238.

[18] S. Jaffard, *Exposants de Hölder en des points donnés et coefficients d'ondelettes,* C. R. Acad. Sci. Sér. I Math., vol. 308, pp. 79–81 (1989).

[19] S. Jaffard, *Multifractal formalism for functions,* SIAM J. Math. Anal., vol. 28, pp. 944–998 (1997).

[20] S. Jaffard, *Wavelet techniques in multifractal analysis,* Fractal Geometry and Applications: A Jubilee of Benoît Mandelbrot, M. Lapidus et M.

van Frankenhuijsen Eds., Proceedings of Symposia in Pure Mathematics, AMS, Vol. 72 Part 2, pp. 91–152 (2004).

[21] S. Jaffard, *Wavelet Techniques for pointwise regularity,* Ann. Fac. Sci. Toul., Vol. 15 no. 1, pp. 3–33 (2006).

[22] S. Jaffard, *Pointwise regularity associated with function spaces and multifractal analysis,* Banach Center Pub. Vol. 72 Approximation and Probability, T. Figiel and A. Kamont Eds., pp. 93–110 (2006).

[23] S. Jaffard, *Hölder width and directional smoothness of nonharmonic Fourier series,* preprint (2007).

[24] S. Jaffard and C. Melot, *Wavelet analysis of fractal Boundaries, Part 1: Local regularity and Part 2: Multifractal formalism,* Comm. Math. Phys. Vol. 258 no. 3, pp. 513–539 (2005).

[25] S. Jaffard and Y. Meyer, *Wavelet methods for pointwise regularity and local oscillations of functions,* Mem. Amer. Math. Soc., vol. 123 No. 587 (1996).

[26] S. Jaffard, Y. Meyer and R. Ryan, *Wavelets: Tools for Science and Technology,* SIAM, (2001).

[27] B. Lashermes, S. Roux et P. Abry and S. Jaffard, *Comprehensive multifractal analysis of turbulent velocity using wavelet leaders,* preprint (2007).

[28] J. Lévy-Véhel and S. Seuret, *The local Hölder function of a continuous function,* preprint (2001).

[29] S. Mallat, *A Wavelet Tour of Signal Processing,* Academic Press (1998).

[30] B. Mandelbrot, *Intermittent turbulence in selfsimilar cascades: divergence of high moments and dimension of the carrier,* J. Fluid Mech., vol. 62, pp. 331–358 (1974).

[31] B. Mandelbrot, *The Fractal Geometry of Nature,* W. H. Freeman (1982).

[32] B. Mandelbrot, *A multifractal walk down Wall Street,* Scienttific American, vol. 280, pp. 70–73 (1999).

[33] P. Mattila, *Geometry of sets and measures in Euclidean Spaces,* Cambridge Univ. Press (1995).

[34] Y. Meyer, *Ondelettes et Opérateurs,* Hermann (1990).

[35] S. Orey and S. J. Taylor, *How often on a Brownian path does the law of iterated logarithm fail?* Proc. London Math. Soc. (3) 28, pp. 174–192.

[36] K. I. Oskolkov, *The Schrödinger density and the Talbot effect,* Approximation and Probability, Banach Center Pub. Vol. 72, pp. 19–219 (2006).

[37] G. Parisi and U. Frisch, *On the singularity structure of fully developed turbulence;* appendix to *Fully developed turbulence and intermittency,* by U. Frisch; Proc. Int. Summer school Phys. Enrico Fermi, 84–88 North Holland (1985).

[38] I. M. Singer, *Bases in Banach spaces,* Vol. 1, Springer (1970).

[39] C. Tricot, *Two definitions of fractional dimension*, Math. Proc. Cambridge Philos. Soc, 91 (1), pp. 57–74 (1982).

[40] C. Tricot, *Function norms and franctal dimensions,* SIAM J. Math. Anal. Vol 28 no. 1, pp. 189–212 (1997).

[41] H. Wendt, P. Abry and S. Jaffard, *Bootstrap for Emperical Multifractal Analysis,* IEEE Signal Proc. Mag., vol. 24, no. 4, pp. 38–48 (2007).

[42] H. Wendt, S. Roux, P. Abry and S. Jaffard, *Wavelet leaders and bootstrap for multifractal analysis of images,* Signal Proces., vol. 89, pp. 1100–1114 (2009).

Wavelet Methods for Image-Based Face Recognition: A Survey*

Chaochun Liu   Daoqing Dai

Center of Computer Vision & Department of Mathematics

Sun Yat-Sen (Zhongshan) University, China

Email: liuchch@mail2.sysu.edu.cn, stsddq@mail.sysu.edu.cn

Abstract

Image-based face recognition is a challenging problem due to variations in pose, expression and illumination. Techniques that can provide effective feature representation with enhanced discriminability are crucial. Due to its attractive attributes, i.e., sparsity, multiresolution, spatial-frequency localization, the wavelet transform has been successfully applied to generate robust discriminant features for face recognition. The goal of this paper is to present a survey of state of the art 2-D face recognition algorithms using wavelet methods. The paper first presents a high-level face recognition framework that reduces the overall process into smaller components. Then applications of wavelets on two most important components: denoising and feature extraction are reviewed. Also, existing problems are covered and possible solutions are suggested.

1 Introduction

Multiscale processing is intuitionistic to develop efficient image recognition algorithms. Many current image coding methods incorporate properties of the human visual system into the data representation, while psychophysics and physiological experiments [1] have shown that multiscale transforms seem to appear in the visual cortex of mammals. This is an important motivation to further study the application of such transforms to image analysis. So the multiscale wavelet transform has received very significant impact on computer vision and image processing, e.g.,

*This project is supported in part by NSF of China (60575004, 10771220), NSF of Guangdong (05101817) and the Ministry of Education of China (NCET-04-0791, SRFDP-20070558043). Daoqing Dai is the corresponding author.

human face recognition. "The wavelet transform appears to be part of the lower levels of the human visual system. Small amplitude high-frequency wavelets are ubiquitous in animal visual systems. They have been observed in the functioning of the human retina, and in the lowest level processing performed in the human visual cortex" [2, 3]. Studies have also indicated that the receptive fields in the visual cortex of cats [4] and other mammals have properties that resemble the Gabor wavelets which provide optimal localization in both the spatial and frequency domains [3, 5]. Therefore, it should be no surprise that researchers are applying wavelets to human face recognition systems.

Although wavelets have attracted much attention in the human face recognition community, there has been no comprehensive review of wavelet applications in the field. In this paper we attempt to fill the void by presenting the necessary foundations for understanding face recognition task and wavelet theory, and a summary of researches on face recognition using wavelets. The reader should be cautioned that our overview may be a little eclectic. An interested reader is encouraged to consult with other papers for further reading.

This paper is organized as follows: Section 2 presents an overview of the human face recognition task, while the structure of a pattern recognition system is briefly shown in Section 3. In Section 4, we introduce some necessary background related to wavelets. Then wavelet applications in human face recognition, including preprocessing, feature extraction will be reviewed in Sections 5, 6. Finally, Section 7 makes a conclusion and discusses future research directions.

2 Face recognition task

Machine recognition of human faces from still and video images has been an active research area spanning several disciplines such as image processing, pattern recognition, computer vision and neural networks [6–8]. "A general statement of the problem can be formulated as follows: given still or video images of a scene, identify or verify one or more persons in the scene using a stored database of faces. Available collateral information such as race, age, gender, facial expression, may be used in narrowing the search. The solution to the problem involves detection and segmentation of faces (face detection) from cluttered scenes, extraction of features from the face regions, verification, or recognition" [6, 8], as shown in Figure 2.1.

Generally, the input digital image/video can be captured by scanner or some photographic equipments, such as vidicon, digital camera, *et al.* "Various applications of face recognition range from static controlled format photographs to uncontrolled video images posing a wide range

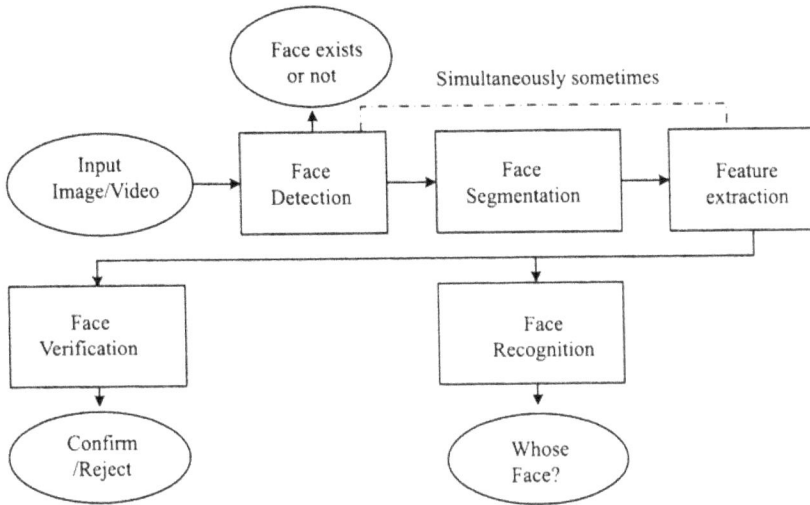

Figure 2.1   Configuration of a generic face recognition system.

of different technical challenges and requiring an equally wide range of techniques from image processing, analysis, understanding, and pattern recognition. One can broadly classify face recognition systems into two groups depending on whether they make use of static images or of video. Within these groups, significant differences exist, depending on the specific application. The differences are in terms of image quality, amount of background clutter (posing challenges to segmentation algorithms), variability of the images of a particular individual that must be recognized, availability of a well-defined recognition or matching criterion, and the nature, type, and amount of input from a user" [6, 8]. In this survey, we focus on *the image-based input*. Various applications also generate different outputs of system, including

- Face detection: verify that is there a face in the input image/video, and detect the place of the face. To evaluate the detection performance, two statistics are important: true positives (also referred to as detection rate) and false positives (reported detections in non-face regions). An ideal system would have very high true positive and very low false positive rates.

- Face verification: one-to-one match that compares a query face image against a template face image whose identity is being claimed. To evaluate the verification performance, the verification rate (the rate at which legitimate users are granted access) vs. false accept rate (the rate at which imposters are granted access) is computed, called ROC curve. A good verification system should balance these two rates based on different security needs [9].

- Face recognition: one-to-many matching process that compares a query face image against all the template images of different persons in a face database to determine the identity of the query face. The query face is assigned to the person whose template image has the highest similarity with the query face. To evaluate the recognition performance, usually the recognition rate is used, which is the ratio of the number of successful recognition and the total number of query face. Sometimes, the "Cumulative Match Score (CMS)" curve is also used [9]. A good recognition system should have high recognition rate or top CMS curve.

2.1 Why face recognition is studied

Over the past 30 years, face recognition has received significant attention. There are at least two reasons for this trend [8]. The first is the wide range of commercial and law enforcement applications, such as

- Information security (identity authentication): personal device logon, database security, internet access, secure trading terminals.
- Public security: crowd surveillance, suspect tracking and investigation, mug shots matching, portal control, postevent analysis, advanced video surveillance.
- Smart cards: drivers' licenses, entitlement programs, immigration, national ID, passports, voter registration.
- Entertainment: video game, virtual reality, training programs, human-robot-interaction, human-computer-interaction.

The second is the availability of feasible technologies after 30 years of research. The common interest among researchers working in diverse fields is motivated by our remarkable ability to recognize people and the fact that human activity is a primary concern both in everyday life and in cyberspace. Researchers in computer vision hope that it can instead of the visual function of human eyes. Researchers in artificial intelligence hope that it can imitate human intelligence. However, face recognition is a most challenging problem across so many disciplines. It is the most typical and difficult problem in pattern recognition.

Although very reliable methods of biometric personal identification exist, such as fingerprint analysis, retinal/iris scan, hand scan, voice scan, signature scan, these methods rely on the cooperation of the participants [10, 11]. Whereas a personal identification system based on analysis of frontal or profile images of the face does not claim for the participant's cooperation or knowledge, is less intrusive than all other technologies and has thus a higher level of user acceptance. "It can offer a user-friendly system that can secure our assets and protect our privacy without losing our identity in a sea of numbers" [8]. So facial scan is an effective and convenience biometric indicator. Among the six biometric indicators considered in [10], facial indicator scores the highest compatibility in a machine readable travel documents system based on a number of evaluation factors [10], as shown in Figure 2.2. But for machine identification, it poses more technological challenges, currently has lower accuracy rates than the other principal indicator. "Face recognition also holds the risk that the biometric identifier may be "stolen" without a person's knowledge as people nearly always have their faces on public display, thus it is critically important to make systems which are practically impossible to spoof" [12].

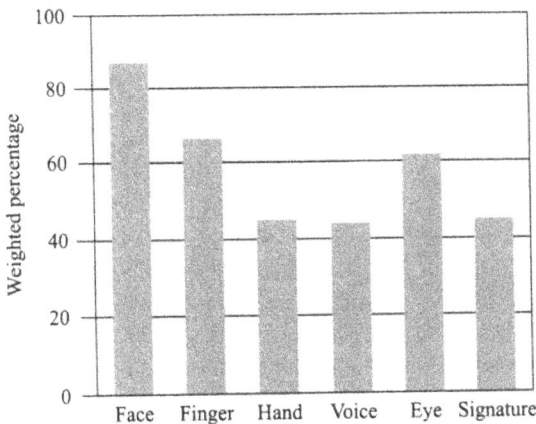

Figure 2.2   Comparison of various biometric features based on MRTD compatibility [10].

2.2   Challenge of face recognition

Face recognition evaluation reports [13] and other independent studies indicate that the bottleneck of face recognition is wide-range variations of human faces, due to pose, occlusion, illumination, and expression which result in a highly nonlinear and nonconvex problem and other factors [6,8]. On the other hand, it is impractical to collect sufficient prototype images covering all the possible variations. Therefore, the challenging research issue is how to construct a small-size-training face recognizer robust to environmental variations. The key technical challenges are summarized below:

- There are large variability/deformation in facial appearance. The

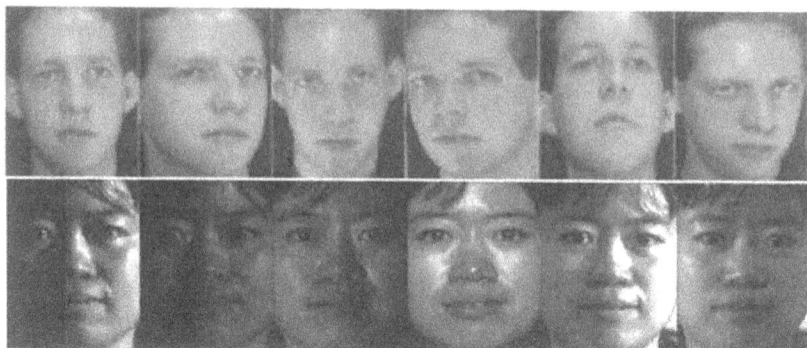

Figure 2.3   Top: Intrasubject variations mainly in pose and expression, the images are from the ORL database. Bottom: Intrasubject variations mainly in illumination, the images are from the CMU Pose, Illumination, and Expression (PIE) database.

shape and reflectance are two intrinsic properties of a face object. The appearance-view of a face is also subjected to several other factors, including the facial pose, expression and illumination. Figure 2.3 shows an example of significant intrasubject variations caused by these factors. In addition to these, "various imaging parameters, such as aperture, exposure time, lens aberrations, and sensor spectral response also increase intrasubject variations. All these factors are confounded in the image data" [14], so "the variations between the images of the same face due to illumination and viewing direction are almost always larger than the image variation due to change in face identity" [15]. This variability makes it difficult to extract the intrinsic information of the face from their respective images.

- Extrinsic factors such as age, hair, face-painting, occlusion greatly complicate the identification of humans. Moreover, the characteristics of humans may be different for different races, e.g., skin color, make the recognition of faces more difficult.

- The wide-range variations of human faces, due to intrinsic and extrinsic factor, result in a highly nonlinear and nonconvex problem. This crucial fact limits the power of the linear methods to achieve highly accurate face recognition, but the nonlinear methods are more complex and immature.

- Face recognition is always a high dimensionality and small sample size ("SSS") problem, because the number of collected prototype images in databases is limited and the dimensionality of pattern space constructed by 2-D facial images is always huge. For example, a canonical face image of 112×92 resides in a 10304-

dimensional pattern space. Nevertheless, the number of examples per person available for training is usually much smaller than the dimensionality of the pattern space. So a system trained on so few samples may not generalize well to unseen instances of the face [14].

2.3   History of face recognition

Over the past 30 years, face recognition has attracted researchers who have different backgrounds: psychology, neurology, pattern recognition, neural networks, computer vision, and computer graphics. Lots of methods on face recognition have been proposed. These research are generally conducted by psychophysicists, neuroscientists and engineers on various aspects of face recognition by humans and machines. "Psychophysicists and neuroscientists have been concerned with issues such as: uniqueness of faces; whether face recognition is done holistically or by local feature analysis; analysis and use of facial expressions for recognition; how infants perceive faces; organization of memory for faces; inability to accurately recognize inverted faces; existence of a "grandmother" neuron for face recognition; role of the right hemisphere of the brain in face perception; and inability to recognize faces due to conditions such as prosopagnosia. Some of the theories put forward to explain the observed experimental results are contradictory. Many of the hypotheses and theories put forward by researchers in these disciplines have been based on rather small sets of images" [6,8]. The earliest work on face recognition can be traced back at least to the 1950s in psychology [16] and to the 1960s in the engineering literature [17]. Lots of studies have important consequences for engineers who design algorithms and systems for machine recognition of human faces, which roughly follow a trace from local to holistical feature analysis.

- Local feature analysis: During the early and mid-1970s, typical pattern classification techniques, which use geometric/statistic attributes of local features in faces or face profiles, were developed. The geometric/statistic attributes are usually the distances/angles between important points or shape constructed by important points. The local features are usually the eyes, nose, and mouth. For example, the width of the head, the distances between the eyes and from the eyes to the mouth are used in [18], and the distances and angles between eye corners, mouth extrema, nostrils, and chin top are used in [19].

- Holistic feature analysis: During the 1990s and present, many methods use the whole face region as the raw input to a recognition system. One kind of these popular methods are statistics-based methods which search an optimal discriminant subspace. For example, the eigenface was developed using principle component analysis (PCA) [20], the fisherface is developed using linear/Fisher discriminant analysis (LDA/FDA) [21–23], and a lot variants of LDA, e.g., generalized discriminant analysis (GDA) [24], regularized discriminant analysis (RDA) [25] were developed subsequently. Another popular kind is wavelet-based methods which decompose facial images into different frequency ranges and prune away the variable subbands containing intrinsic deformations due to expression or extrinsic factors (like illumination, occlusions), the most relevant information is retained to better represent the data.

- Hybrid methods: With the development of holistic-feature-based methods, some studies combine both local features and whole face region to recognize a face, just as the human perception system. These methods may potentially offer the best of the two types of methods. For example, the use of hybrid features by combining eigenfaces and other eigenmodules: eigeneyes, eigenmouth, and eigennose, is explored in [26]. A hybrid representation based on PCA and local feature analysis (LFA) [27], and a flexible appearance model-based method [28] were also developed. Someone argue that these types of methods are important and deserve further investigation. Perhaps many relevant problems need to be solved before fruitful results can be expected, for example, how to optimally arbitrate the use of holistic and local features [8].

3   The structure of a Pattern Recognition System (PRS)

Face recognition is considered as the most typical and difficult problem in pattern recognition. We will try to shed some light on this problem and the techniques that have been brought to bear on its attempted solution. Considering the strategy: divide then conquer separately while encountering a difficult problem, it is instructive to divide into several intermediate stages, the process of starting with a pattern in the physical world and ending with a decision of class-membership of that pattern. Figure 3.1 introduces the basic building blocks of a typical pattern recognition system.

Generally it is more easy and effective to solve each sub-problem optimally. However, we must remember that ultimately all the solutions of these sub-problems must be woven back together again and the optimality of the entire system is only as optimal as the weakest link in the chain [29]. Let us consider these sub-problems in more detail [29, 30].

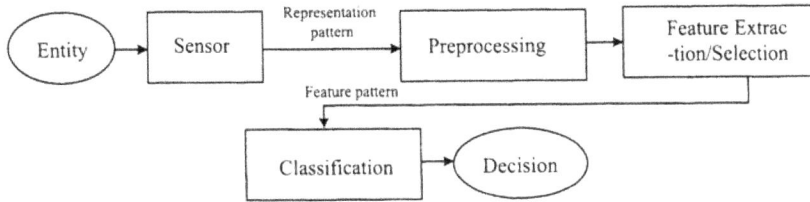

Figure 3.1   Basic building blocks of a typical pattern recognition system.

- Sensor: A sensor converts an entity (e.g., human face) from the "real" physical world, into a digital pattern in the computer. 2-D digital pattern we usually call a digital image. It consists usually of a square array or matrix of numbers. Each element or cell in the array of a digital image is called a pixel and the number associated with a pixel represents the light intensity at that location in the image.

- Preprocessing: Generally preprocessing has three categorization: the first category is denoising, because it may be the case that the sensor is sensitive to some extrinsic factors in the input field of vision that cause random changes of pixels, referred to as noises. A family of procedures for smoothing, enhancing, filtering, cleaning-up and otherwise massaging a digital image is necessary so that subsequent algorithms along the road to final classification can be made simple and more accurate. The second category is detecting and segmenting, that is detecting and segmenting the interest meaningful objects from an image, so that the uninterested parts have no effect on final classification. The third is scaling, sometimes, scaling all data into an uniform measure is necessary.

- Feature extraction/selection: For the reason of easier subsequent analysis, improved classification performance through a more stable representation, removal of redundant or irrelevant information or an attempt to discover underlying structure by obtaining a graphical representation, feature extraction/selection, which is the name given to a family of procedures for measuring the relevant intrinsic information contained in a pattern, is usually the most important step in whole PRS. The procession can be described as a transform from original pattern space to feature space, which can be linear or nonlinear. Generally, the dimensionality of the original pattern space is bigger than the dimensionality of feature space, that is, the transform is commonly used for reducing the dimensionality of data so that the extracted/selected features are as representative as possible by the sense of certain optimal criterion. Feature extraction is to recover new meaningful underlying variables or features that describe the data, by a transformation which may be a linear/nonlinear combination of the original variables. Feature selection is to identify those variables that contribute to the classification task and discard those variables that do not contribute to class separability. Thus, the procedure is to seek a optimal feature subset out of the original variable ensemble. It should be noted that there is no theory either in computer science or psychology to solve this sub-problem optimally.

- Classification: Concern with classifying the pattern into one or more categories on which some subsequent task depends, then a pattern class can be represented by a region or sub-space of the feature space. That is, classification becomes a matter of determining in what region of the feature space an unknown pattern falls into. The task in any given situation is to design a decision rule that is easy to compute and will minimize the probability of misclassification relative to the power of the feature extraction scheme employed. A classifier can be designed using a number of possible approaches. The simplest and the most intuitive approach is based on the concept of similarity, i.e., patterns that are similar should be assigned to the same class. The second common concept used for designing classifiers is probabilistic-based. The third category of classifiers is to construct decision boundaries directly by optimizing certain error criterion. In practice, the choice of a classifier is a difficult problem and it is often based on which classifier happens to be available, or best known to the user.

Generally speaking, the wavelet transform is a tool that divides data, functions, or operators into different frequency components and then studies each component with a resolution matched to its scale [31]. Wavelets have many favorable properties, such as vanishing moments, hierarchical and multiresolution decomposition structure, linear time and space complexity of the transformations, decorrelated coefficients, and a wide variety of basis functions. These properties could provide considerably more efficient and effective solutions to many pattern recognition problems. Therefore, the wavelet transform is used to provide economical and informative mathematical representation of many pattern of interest [32] and of very practical use in pattern recognition task.

In practice, a wide variety of wavelet-related methods have been mainly applied to two mid-steps of pattern recognition problems. We will pay our attentions to the wavelet applications on the mid-steps—image denoising preprocession and feature extraction/selection for face recognition in Sections 5 and 6.

4   Wavelet background

Wavelets are functions that satisfy certain mathematical requirements and are used in presenting data or other functions, similar to sines and cosines in the Fourier transform. However, they represent data at different scales or resolutions, which distinguishes the wavelet transform from the Fourier transform. In this section, we will present the basic foundations that are necessary to understand and use wavelets.

4.1   The one dimensional wavelet transform

In real applications, the discrete wavelet transform (DWT) is usually used to represents a one-dimensional (1-D) signal $f(t)$ in terms of shifted versions of a lowpass real function $\phi(t)$ and shifted and dilated versions of a prototype bandpass real function $\psi(t)$, where $\phi(t)$ is called scaling function, and $\psi(t)$ is called mother wavelet function whose dilated and translated family

$$\left\{\psi_{j,k}(t) = 2^{-j/2}\psi(2^{-j}t - k), j, k \in \mathbb{Z}\right\}$$

form an orthonormal basis of $L^2(\mathbb{R})$ [31]. These basis functions are usually referred to as wavelets. The function $\psi(t)$ is assumed to satisfy the admissibility condition:

$$C_\psi = \int_\mathbb{R} \frac{|\Psi(\omega)|^2}{|\omega|} d\omega < \infty \qquad (4.1)$$

where $\Psi(w)$ is the Fourier transformation of $\psi(t)$. The admissibility condition (4.1) implies $\Psi(0) = \int \psi(t)dt = 0$ which refers to oscillatory, motivating the name wavelet. The "diminutive" appellation comes from the fact that ψ can be of finite length or compactly supported and well localized with arbitrarily fine by appropriate scaling. Then we can represent $f(t)$ as

$$f = \sum_k c_{j_0,k}\phi_{j_0,k} + \sum_{j \leq j_0} \sum_k d_{j,k}\psi_{j,k} \quad \text{with}$$

$$c_{j_0,k} = \langle f, \phi_{j_0,k}\rangle = \int_\mathbb{R} f(t)\phi_{j_0,k}(t)dt, \ d_{j,k} = \langle f, \psi_{j,k}\rangle = \int_\mathbb{R} f(t)\psi_{j,k}(t)dt.$$
$$(4.2)$$

The scaling coefficient $c_{j_0,k}$ measures the local low frequency content around time $2^{j_0}k$. The wavelet coefficient $d_{j,k}$ measures the local high frequency content around time $2^j k$. (4.2) employs scaling coefficients only at scale j_0 and wavelet coefficients at resolutions $j \leq j_0$, which add higher resolution details to the signal. The partial sum of wavelet

$\sum_{k=-\infty}^{+\infty} \langle f, \psi_{j,k} \rangle \psi_{j,k}$ $(j \leq j_0)$ can be interpreted as the approximation at the resolution j. The approximations of signals at various resolutions with orthogonal projections can be computed by multiresolution which is characterized by particular discrete filters that govern the loss of information across resolutions. These discrete filters provide a simple procedure called pyramid algorithm for decomposing and synthesizing wavelet coefficients at different resolutions [33]:

$$c_{j+1,k} = \sum_l \overline{h_{l-2k}} c_{j,l}, \quad d_{j+1,k} = \sum_l \overline{g_{l-2k}} c_{j,l},$$

$$c_{j,k} = \sum_l (h_{k-2l} c_{j+1,l} + g_{k-2l} d_{j+1,l})$$

where h_k, g_k are discrete filter sequences, they satisfy respectively

$$\phi(t) = \sum_k h_k \phi(2t - k), \quad \psi(t) = \sum_k g_k \phi(2t - k), \quad g_k = (-1)^k \overline{h_{1-k}}.$$

The two-channel filter bank method parallelly filters a signal by the low-pass filter h and highpass filter g followed by subsampling. The filter h removes the high frequencies and retains the low frequency components, the filter g removes the low frequencies and produces high frequency components. Together, they decompose the signal into different frequency subbands, and downsampling is used to kept half of the output components of each filter. For the wavelet transform, only the lowpass filtered subband is further decomposed.

4.2   The two-dimensional wavelet transform

The two-dimensional (2-D) wavelet can be constructed from the tensor product of one-dimensional ϕ and ψ by setting:

$$\begin{aligned} \phi(x,y) &= \phi(x)\phi(y), \quad \psi^H(x,y) = \psi(x)\phi(y), \\ \psi^V(x,y) &= \phi(x)\psi(y), \quad \psi^D(x,y) = \psi(x)\psi(y) \end{aligned} \tag{4.3}$$

where $\psi^H(x,y), \psi^V(x,y), \psi^D(x,y)$ are 2-D wavelet functions. Their dilated and translated family

$$\{\psi^\lambda_{j,k_1,k_2}(x,y) : j, k_1, k_2 \in \mathbb{Z}, \lambda = H, V, D\}$$

and

$$\{\phi_{j,k_1,k_2}(x,y) : j, k_1, k_2 \in \mathbb{Z}\}$$

form an orthonormal basis of $L^2(\mathbb{R}^2)$. For every $f \in L^2(\mathbb{R}^2)$, it can be represented as

$$f = \sum_{k \in \mathbb{Z}^2} c_{j_0,k} \phi_{j_0,k} + \sum_{j \leq j_0, k \in \mathbb{Z}^2, \lambda = H, V, D} d^\lambda_{j,k} \psi^\lambda_{j,k},$$

$$c_{j_0,k} = \langle f, \phi_{j_0,k} \rangle, \quad d^\lambda_{j,k} = \langle f, \psi^\lambda_{j,k} \rangle.$$

Similar to 1-D wavelet transform of signal, in image processing, the approximations of images at various resolutions with orthogonal projections can also be computed by multiresolution which is characterized by the two-channel filter bank that governs the loss of information across resolutions. The 1-D wavelet decomposition is first applied along the rows of the images, then their results are further decomposed along the columns. This results in four decomposed subimages L_1, H_1, V_1, D_1. These subimages represent different frequency localizations of the original image which refer to Low-Low, High-Low, Low-High and High-High respectively. Their frequency components comprise the original frequency components but now in distinct ranges. In each iterative step, only the subimage L_1 is further decomposed. Figure 4.1 (Top) shows a two-dimensional example of facial image for the wavelet decomposition with depth 2.

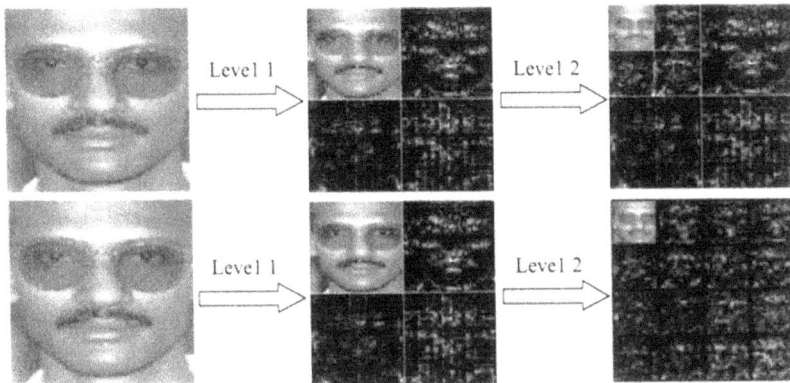

Figure 4.1   Top: The two-dimensional wavelet decomposition of facial image with depth 2. Bottom: The two-dimensional wavelet packet decomposition of facial image with depth 2.

4.3   The wavelet-packet transform

There are complex natural images with various types of spatial-frequency structures, which motivates the adaptive bases that are adaptable to the variations of spatial-frequency. Coifman and Meyer [34] introduced an orthonormal multiresolution analysis which leads to a multitude of orthonormal wavelet-like bases known as wavelet packets. They are linear combinations of wavelet functions and represent a powerful generalization of standard orthonormal wavelet bases. Wavelet bases are one particular version of bases that represent piecewise smooth images effectively. Other bases are constructed to approximate various-type images of different spatial-frequency structures [33].

As a generalization of the wavelet transform, the wavelet packet co-efficients also can be computed with two-channel filter bank algorithm. The two-channel filter bank is iterated over both the lowpass and high-pass branches in the wavelet packet decomposition. Not only L_1 is fur-ther decomposed as in the wavelet decomposition, but also H_1, V_1, D_1 are further decomposed. This provides a quad-tree structure correspond-ing to a library of wavelet packet basis, and images are decomposed into both spatial and frequency subbands, as shown in Figure 4.1 (bottom).

4.4   Properties of wavelets for image processing

Generally, local image contrasts are often more informative than light intensity values. The wavelet transform can be interpreted as a multi-scale differentiator or variation detector that represents the singularity of an image at multiple scales and three different orientations—horizontal, vertical, and diagonal. Contours of image structures correspond to sharp contrasts (singularity) are represented by a cascade of large wavelet co-efficients across scale [33]. If the singularity is within the support of a wavelet basis function, then the corresponding wavelet coefficient is large. Contrarily, the smooth image region is represented by a cascade of small wavelet coefficients across scale. These properties endow us with good ability to recognize complex scenes from a drawing that outlines edges. So it is believed that the wavelet theory relates the behavior of multiscale edges to local image properties.

Some researchers have studied several features of the wavelet trans-form for natural images [33, 35–38]:

- Multiresolution: It provides a simple hierarchical framework with different scales or resolutions for interpretating the image informa-tion [39]. At different resolutions, the details of an image generally characterize different physical structures of the scene. At a coarse resolution, these details correspond to the larger structures which provide the image content. It is therefore natural to analyze first the image details at a coarse resolution and then gradually increase the resolution [40]. Such a coarse-to-fine strategy is useful for face recognition algorithms.

- Compact support: Each wavelet basis function is supported on a finite interval, which guarantees the localization of wavelets, that is, processing a region of image with wavelet does not affect the data out of this region. So that the image can be decomposed into subbands that are localized in both space and frequency domains [41].

- Vanishing moments: With higher vanishing moments, if data can be represented by low-degree polynomials, their wavelet coefficients

are equal to zero, which leads to sparsity property: a wavelet coefficient is large only if the singularities are present in the support
of a wavelet basis function. The magnitudes of coefficients tend to
decay exponentially across scale. So most energy of image concen-
trates on those large coefficients.

- Decorrelation: Wavelet coefficients of images tend to be approximately decorrelated because of the orthonormal property of wavelet
  basis functions. Hence, the wavelet transform could be able used
  to reduce the complexity in the spatial domain into a much simpler
  process in the wavelet domain.

- Parseval's theorem: The energy of image is preserved under the
  orthonormal wavelet transform. Hence the distances between any
  two patterns are not changed by the transform.

- Computation complexity: The computation of the wavelet transform based on pyramidal algorithm can be very efficient. It only
  needs $O(N)$ multiplications. The space complexity is also linear.

So the wavelet transform provides a way to estimate the underlying
function from the data. With the sparsity property, we know that only
some wavelet coefficients are significant in most cases. By retaining selec-
tive wavelet coefficients, the wavelet transform could then be applied to
denoising preprocession and feature extraction/selection (usually means
dimensionality reduction).

5   Preprocessing: wavelets for noise removal

Real world entities are usually not directly suitable for performing pattern recognition algorithms [42]. They contain noise, missing values and
may be inconsistent. Therefore, we usually need preprocessing to remove noise before recognition tasks. In this section, we will elaborate
the applications of wavelets in denoising.

Denoising is an important step in the analysis of images, wavelet techniques which provide an effective way to denoise, have been successfully
applied in image research [43–46]. The wavelet transform employs both
low-pass and high-pass filters to the data. Generally, the low-frequency
parts reflect the data information, and the high-frequency parts are considered as the noise and the data details, so a compromise has to be made
between noise reduction and preserving significant data details in denoising. In fact, the noisy data can usually be approximated by low-degree
polynomial if the data are smooth in most of regions, their corresponding wavelet coefficients are usually small due to vanishing moments of
the wavelet. Therefore, thresholding to the decomposed high-frequency
coefficients on each level can effectively denoise the data [33, 43, 45, 47].

For simplicity, we consider the 1-D signal. Suppose the real signal $f[n]$ of size N is contaminated by the addition of a noise $\epsilon[n]$. It is commonly assumed that $\epsilon[n]$ is independent from the signal and is independent and identically distributed (iid) Gaussian random variables. The observed signal is

$$\hat{f}[n] = f[n] + \epsilon[n], \; n = 0, \cdots, N-1.$$

The signal f is estimated by transforming the noisy data $\hat{f}$ with a decision operator Q and the resulting estimator (denoised signal) is $\tilde{f} = Q\hat{f}$. The goal is to minimize the error of the estimation, which is measured by a loss function. The mean square error (MSE) is a familiar loss function, the MSE of the estimator $\tilde{f}$ of f is the average loss:

$$MSE(Q, f) = E\{\|f - \tilde{f}\|^2\} = E\{\|f - Q\hat{f}\|^2\}.$$

The main idea of wavelet denoising is to transform the noisy data into a wavelet basis, where the large coefficients are mainly the useful information and the smaller ones represent noise. By suitably modifying the coefficients in the new basis, noise can be directly removed from the data. The most popular methodology for estimating f is called waveShrink [43, 47], which includes three steps:

- Wavelet decomposition: Transform the noisy data $\hat{f}$ into wavelet domain. Suppose $\hat{f}$ is decomposed in a wavelet basis $B = \{b_m, 1 \le m \le N\}$, then the wavelet coefficients of $\hat{f}$ can be represented as

$$\langle \hat{f}, b_m \rangle = \langle f, b_m \rangle + \langle \epsilon, b_m \rangle, \; 1 \le m \le N.$$

- Wavelet thresholding: Apply some thresholding methods to the wavelet coefficients, thereby shrink those coefficients smaller than certain amplitude towards zero. There are two commonly used shrinkage functions, the first one is the hard thresholding estimator:

$$\rho_T(x) = \begin{cases} x, & \text{if } |x| > T, \\ 0, & \text{if } |x| \le T, \end{cases}$$

the second one is the soft thresholding estimator:

$$\rho_T(x) = \begin{cases} x - T, & \text{if } x > T, \\ x + T, & \text{if } x < -T, \\ 0, & \text{if } |x| \le T. \end{cases}$$

The threshold T is generally chosen so that there is a high probability that it is just above the maximum level of the noise, $T = \sigma\sqrt{2\ln N}$ is proposed by Donoho and Johnstone [47] on the meaning of minimizing risk $MSE(Q, f)$. In 2-D case, the image $f[n_1, n_2]$

contaminated by a white noise is decomposed in a separable two-dimensional wavelet basis, Figure 5.1 illustrates an example of image denoising using the hard and soft thresholding in the Symmlet 4 wavelet basis.

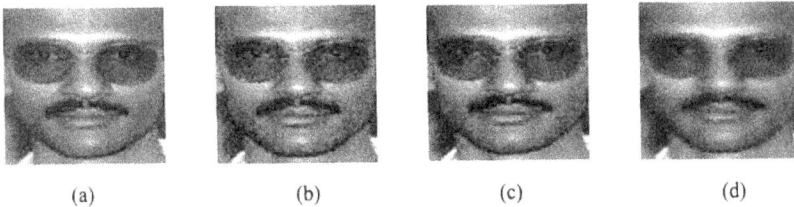

(a)	(b)	(c)	(d)

Figure 5.1   (a) Original image, (b) Noisy image (SNR = 19.95), (c) Estimation with a hard thresholding in a separable wavelet basis (Symmlet 4) (SNR = 22.03), (d) Soft thresholding (SNR = 19.96).

- Reconstruction: Transform the shrunk coefficients back to the data domain. That is, the thresholding estimator $\tilde{f}$ in the wavelet basis $B = \{b_m, 1 \leq m \leq N\}$ can be written as

$$\tilde{f} = Q\hat{f} = \sum_{m=1}^{N} \rho_T(\langle \hat{f}, b_m \rangle) b_m.$$

This process is essentially a form of non-parametric and nonlinear regression, due to the thresholding step [48]. In the whole process, a suitable wavelet, an optimal decomposition level for the hierarchy, one appropriate thresholding function and optimal threshold should be considered [33]. It should be noted that thresholding noisy wavelet coefficients can create small ripples near discontinuities. Indeed, setting a coefficient to zero will introduce oscillations whenever the coefficient is non-negligible, which reduces the signal-noise-ratio (SNR). A method to attenuate the oscillations is to introduce a shift invariant estimation. So design of a shift invariant estimation based on wavelets is still demanded.

6   Wavelet for feature extraction

Feature extraction in the sense of some linear or nonlinear transform of the data with subsequent feature selection is commonly used for reducing the dimensionality of facial image so that the extracted feature is as representative as possible. The images may be represented by their original spatial representation or by frequency domain coefficients. Features that are not obviously present in one domain may become obvious in the other domain. Unfortunately, Heisenberg uncertainty theorem implies that the information can not be compact in both spatial and frequency domain simultaneously. So, neither approach is ideally suited for all kinds of feature distribution. It motivates the use of the wavelet transform which represents both the spatial and frequency domain simultaneously [3]. Also, local frequency information at individual spatial locations alone would be insufficient for representing and detecting features across scales, so the wavelet transform which possesses multiresolution property is more appropriate to represent and extract features across different scales. In our knowledge, existing researches of the methodology mainly focus on the critical-sampled separable 2-D DWT and Gabor wavelet transform. Since a detail review on Gabor wavelets for face recognition has been presented in [49]. In this survey, we will fix attention on the critical-sampled separable 2-D DWT for feature extraction. In addition, we will present a new technique for feature extraction of facial image [50], which is the dual-tree complex wavelet transform ($\mathbb{C}$WT).

6.1   Feature extraction based on the separable wavelet transform

For face recognition, the separable wavelet transform have been popularly used in facial image processing. Its ability to capture localized spatial-frequency information of image motivates its use for feature extraction. The decomposition of the data into different frequency ranges allows us to isolate the frequency components introduced by intrinsic deformations due to expression or extrinsic factors (like illumination) into certain subbands. lots of wavelet-based methods prune away these variable subbands, and focus on the subbands that contain the most relevant information to better represent the data. These methods use the wavelet transform for extracting features commonly in three ways [37]:

- Direct use of wavelet coefficients.
- From combination of wavelet coefficients.
- Searching the best feature in the wavelet packet library.

It should be pointed out that categorizing a specific wavelet technique/ paper into a way of the framework is not strict or unique. Many techniques could be categorized as performing on different ways. We will try to discuss the wavelet techniques with respect to the most relevant way based on our knowledge.

1) Direct use of wavelet coefficients

The simplest application of the wavelet transform for face recognition uses directly wavelet coefficients as features. The wavelet transform can locally detect the multiscale edges of facial images, the lineament edge information exists in the lowest spatial-frequency subband, while finer edge information presents in the higher spatial-frequency subband. After the 3-level separable 2-D wavelet transform, the third-level lowest spatial-frequency subimage with a matrix of $(n_{row}/8) \times (n_{col}/8)$ is extracted as the feature vector, referred to as waveletface [51], where $n_{row} \times n_{col}$ is the resolution of facial image. Generally, low frequency components represent the basic figure of an image, which are less sensitive to image variations. These components form the most informative subimage gearing with the highest discriminating power. The waveletface can be expressed by a form of linear transformation: $y = T_{waveletface}x$, where $T_{waveletface}x$ is composed of impulse responses of the low pass filter h. Different from some statistics-based methods, such as eigenface [20] and fisherface [21], the waveletface can be independently extracted without the effect of new enrolled users and more efficient. It only cost $O(n_{row} \times n_{col})$.

2) From combination of wavelet coefficients

The direct use of wavelet coefficients may not extract the most discriminative features for two reasons:

- There is much redundant or irrelevant information contained in wavelet coefficients.

- It can not recover new meaningful underlying features which have more discriminative power.

In order to overcome the deficiency of direct use of wavelet coefficients, it is possible to construct features from the combinations of wavelet coefficients to produce a low-dimensional manifold with minimum loss of information so that the relationships and structure in the data can be identified. These can be done in two ways:

- Use the statistical quantity, energy and entropy of wavelet coefficients in each spatial-frequency subband as discriminative features.

- Employ traditional transforms (e.g., PCA, LDA, independent component analysis (ICA), associate memory (AM), Neural Networks) to extract enhanced discriminative features in one or several special spatial-frequency subbands.

2.a) Use the statistical quantity, energy and entropy as discriminative features

Generally, the statistical moments (e.g., mean, variance), energy and entropy, are usually helpful to represent features or characteristics of data, they are very simple and requires less computation load.

Garcia *et al.* [52, 53] presented a wavelet-based framework for face recognition. Each face is described by a subset of subband images after two-level wavelet packet transform. Then a set of simple statistical measures (mean, variance) of all subbands are computed to characterize face information and reduce dimensionality, which form compact and meaningful feature vectors. Instead of the mean and variance, the Zernike Moments is selected as feature extractor in [54], due to its robustness to image noise, geometrical invariants property and orthogonal property.

In [55], after two-level wavelet packet decomposition, the matries' norm of 16 subimages in the second level are computed, forming a face feature vector and classified by support vector machine (SVM). In [56], for each partitioned subblock in horizontal, vertical and diagonal high frequency subbands at all decomposition level, the total energy of all wavelet coefficients in the subblock is computed, forming a component of the wavelet energy feature (WEF) vector. It reflects the strength of the images texture at different positions, directions and resolutions, and is used for facial expression recognition. In [57], after three-level wavelet transform, wavlet energy entropy (WEE) features of four images including low spatial-frequency components are used for recognition. WEE method captures high order statistics information, and at the same time maintains a fast speed.

In fact, other statistical measures, e.g., other kinds of moments can be used in the above wavelet-based framework for face recognition. Moreover, the discrete density function of whole wavelet coefficients in each subband can also be evaluated for recognition. The similarity measure of density function can be computed by some relative entropy, such as Kullback-Leibler divergence or J-divergence [58].

2.b) Employ traditional transform to extract enhanced discriminative features

Generally, the wavelet coefficients are deficient to be good discriminative features, a further discriminant analysis is necessary for recovering new meaningful underlying features which have more discriminative power. The traditional transforms (e.g., PCA, LDA, ICA, AM, Neural Networks, et al.) are very popular for their simplicity and practicality. They can be performed on one or several special frequency subbands which may be chosen by certain criterion, to extract enhanced features for face recognition. The methodology has became most popular in the field of wavelet methods for face recognition.

We proposed a wavelet subband approach on using PCA for human face recognition [59]. Three-level wavelet transform is adopted to decompose an image into different subbands with different frequency components. A midrange frequency subband is selected for PCA representation. A little differently, PCA is perform on low frequency subband after

2-level DWT in [60,61]. In [62], PCA is performed on four subhand generated by 1-level wavelet transform, then a decision fusion is used to get the last result, similar methodology is shown in [63]. The combination of the wavelet transform and PCA/eigenface is also shown in [64,65].

In [66,67], we analyzed the role of the wavelet transform: low-pass filtering reduces the dimensionality of input data but meanwhile increases the magnitude of the within-class covariance matrix so that the variation information plays too strong a role and the performance of the system will become poorer. So we proposed two wavelet enhanced regularization LDA system for human face recognition, they can adequately utilize the information in the null space of withinclass scatter matrix and solve the small sample size problem encountered in traditional LDA. Similarly, the generalized kernel Fisher discriminant is performed on low frequency wavelet subband in [68].

Ekenel et al. introduced a ternary-architecture multiresolution face recognition system in [69]. They used the 2-D DWT to generate coarse approximations of the face as well as contour, horizontal, vertical and diagonal details of faces at various scales. Subsequently, the PCA or ICA features are extracted from these subbands. They exploited these multiple channels by fusing their information for improved recognition. Similar methodology is shown in [70].

In [71], they proposed a modular face recognition scheme by combining the techniques of wavelet subband representations and kernel associative memories. The kernel associative memory (KAM) model is built up on low frequency wavelet subband for each subject, with the corresponding prototypical images without any counter examples involved. Multiclass face recognition is thus obtained by simply holding these associative memories. Similarly, an adaptive and incremental associative memory is built up on low frequency wavelet subband in [72].

In [73], the DWT is used for feature extraction and dimensional reduction, then a neural network based system is built on low frequency subband for face recognition. In [74], a approach based on wavelet and neural network is proposed for illumination compensation of face image. The method can compensates different scale features of the face image by using the multiresolution characteristic of the wavelet. In [75], the wavelet networks which has the advantage that the wavelet coefficients are directly related to the image data through the wavelet transform is used for face recognition.

Moreover, there are some other techniques employed to extract enhanced discriminative features in wavelet domain in existent researches. In [76] , two techniques based on the multiscale wavalet transform, using "error tolerance" and "region tagging" segmentation algorithms are used for face recognition. In [77], wavelet-based partial representation of a face is used along with locally discriminating projection (LDP), the technique can well enhance the class structure of the data with local and directional information for automatic face recognition task. In [78], they introduced a frequency domain method for performing illumination tolerant face recognition by combining the wavelet decomposition and the quaternion correlation filter techniques [79]. In [80], they developed a framework for recognizing face images by combining the wavelet decomposition, fisherface method, and fuzzy integral. The fisherfaces are generated in four wavelet subband and sent into the fuzzy integral classifier. In [81], the holistic Fourier invariant features are extracted from the low-frequency wavelet subband, referred to as "spectroface" for recognition.

3) Search local discriminant basis/coordinates in wavelet packet library

As a generalization of the wavelet transform, the wavelet packet transform not only offers us an attractive tool for reducing the dimensionality by feature extraction, but also allows us to create localized subbands of the data in both space and frequency domains. A wavelet packet dictionary provides an over-complete set of spatial-frequency localized basis functions onto which the facial images can be projected in a series of subbands. The main design problem for a wavelet packet feature extractor is to choose which subset of basis functions from the dictionary should be used. Most of the wavelet packet dictionary methods that have been proposed in the literature are based on algorithms which were originally designed for signal compression such as the best basis algorithm [82], or the matching pursuit algorithm [83].

Saito and Coifman introduced the local discriminant basis (LDB) algorithm based on a best basis paradigm, searching for the most discriminant subbands (basis) that illuminates the dissimilarities among classes from the wavelet packet dictionary [84, 85]. It first decomposes the data in the wavelet packet dictionary, then data's energies at all coordinates in each subband are accumulated for each class separately to form a spatial-frequency energy distribution per class on the subband. Then the difference among these energy distributions of each subband is measured by a certain "distance" function (e.g., Kullback-Leibler divergence), a complete local discriminant basis (LDB) is selected by the difference-measure function using the best basis algorithm [82], which can represent the distinguishing features among different classes. After the basis is selected, the loadings of their coordinates are fed into a traditional classifier such as LDA or classification tree (CT). Finally, the corresponding coefficients of probes are computed and fed to the classifier to predict their classes.

Unfortunately, the energies may not be so indicative for discrimination sometimes, because not all coordinates in the LDB are powerful to distinguish different subjects. Many less discriminant coordinates may add up to a large discriminability for a subband. An example of artificial problem was used to validate that it may be fail to select the right basis function as a discriminator [86]. So Saito and Coifman suggested a modified version of the LDB (MLDB) algorithm which uses the empirical probability distributions instead of the space-scale energy as their selection strategy to eliminate some less discriminant coordinates in each subband locally [86]. It estimates the probability density of each class in each coordinate in all subbands. Then the discriminative power of each subband is represented by the first N_0 most discriminant coordinates in terms of the "distance" among the corresponding densities. This information is then used for selecting a basis for classification as in the original LDB algorithm. Although the MLDB algorithm may overcome some shortage of LDB, the selection of coordinates is only limited to each subband so that the coordinates in different subbands are still incomparable. Therefore, the MLDB algorithm just gives an alternative to the original LDB.

This LDB concept has become increasingly popular and has been applied to a variety of classification problems. Based on the LDB idea, Kouzani et al. proposed a human face representation and recognition system in [87]. An optimal transform basis, called the face basis, is identified as face feature for face recognition. In [88], after using the wavelet packet decomposition to create localized space-frequency subspaces of the original facial image, PCA is performed in all subbands to determine their figures of merit (FOM) for evaluating the importance of them, then the subbands which generalize better across illumination variations are sought as discriminative features, by using the best basis algorithm.

Since features with good discriminability may locate in different subbands, it is important to find them among all subbands instead of certain specific subbands. We proposed a novel local discriminant coordinates (LDC) method based on wavelet packet to compensate for illumination, pose and expression variations for face recognition [89]. The method searches for the most discriminant coordinates from the wavelet packet dictionary, instead of the most discriminant basis in LDB. The LDC idea makes use of the scattered characteristic of best discriminant features, the feature selection procedure is independent of subbands, and only depends on the discriminability of all coordinates, because any two coordinates in wavelet packet dictionary are comparable for their discriminability which is computed by a maximum a posterior (MAP) logistic model based on a dilation invariant entropy. LDC-based feature extraction not only automatically selects low frequency components, but also middle frequency components whose judicious combination with low frequency components can improve the performance of face recognition greatly.

4) Robust issue of standard separable 2-D wavelet transform for face recognition

It is known that a good feature extractor of face recognition system is claimed to select as more as possible the best discriminative features which are not sensitive to arbitrary environmental variations.

Nastar et al. investigated the relationship between variations in facial appearance and their deformation spectrum [90]. They found that facial expressions and small occlusions affect the intensity manifold locally. Under frequency-based representation, only high-frequency spectrum is affected. Moreover, changes in pose or scale of a face and most illumination variations affect the intensity manifold globally, in which only their low-frequency spectrum is affected. Only a change in face will affect all frequency components.

So there are no special subbands whose all coordinates are not sensitive to these variations. In each subband, there may be only segmental coordinates have enough discriminability to distinguish different person, the remainder may be sensitive to environmental changes. So some methods based on the whole subband may unavoidably use these sensitive features which may affect their performance for face recognition.

Moreover, there may be no special subbands containing all the best discriminant features, because the features not sensitive to environmental variations are always distributed in different coordinates of different subbands locally. So methods based on the segmental subbands may lose some good discriminative features.

Furthermore, in different subbands, the amount and distribution of best discriminant coordinates are always different. Many less discriminant coordinates in one subband may add up to a larger discriminability than another subband whose discriminability is added up with few best discriminant coordinates and residual small discriminant coordinates. The few best discriminant coordinates may be discarded by some methods which search for the best discriminate subbands, but usually only the few best discriminant coordinates are needed. So the best discriminant information selection should be independent of their seated subbands, and only depends on their discriminability for face recognition.

On the other hand, although the the standard 2-D DWT/DWPT have been successfully and popularly used for face recognition, they suffer from four fundamental, intrinsic and intertwined shortcomings:

- Shift variance: Mallat investigated that the wavelet transform lacks translation invariance [91]. It is difficult to characterize a pattern from the wavelet coefficients in a basis since these wavelet descriptors depend upon the pattern location. To simplify the explanation, suppose $f(t)$ is a 1-D signal, $f_\tau(t) = f(t-\tau)$ $(\tau \in \mathbb{Z})$ is the translated version of f by τ. Their inner product with discrete wavelet basis function $\psi_{j,k}$ $(j, k \in \mathbb{Z})$ dilated by 2^j and translated by $2^j k$ is

$$\langle f, \psi_{j,k} \rangle = \int_{-\infty}^{+\infty} f(t) 2^{-\frac{j}{2}} \psi(2^{-j}t - k) \mathrm{d}t \triangleq T_f(j, k),$$

$$\langle f_\tau, \psi_{j,k} \rangle = \int_{-\infty}^{+\infty} f(t - \tau) 2^{-\frac{j}{2}} \psi(2^{-j}t - k) \mathrm{d}t$$

$$= \int_{-\infty}^{+\infty} f(t) 2^{-\frac{j}{2}} \psi(2^{-j}t - k + 2^{-j}\tau) \mathrm{d}t = T_f(j, k - 2^{-j}\tau).$$

However, the translation of $T_f(j, k)$ by τ is

$$(T_f)_\tau (j, k) = T_f(j, k - \tau) \neq T_f(j, k - 2^{-j}\tau) \text{ when } j \neq 0.$$

It implies that the standard wavelet transform is shift variant. As a result, the wavelet coefficients of the translated function $f_\tau(t)$ may be very different from the wavelet coefficients of $f(t)$. For facial image, the wide-range variations, due to pose, illumination, and expression, always cause shifts of patterns, which implies that the standard wavelet transform is not perfect for feature extraction of facial images.

- Oscillations: Since wavelets are bandpass functions, the wavelet coefficients tend to oscillate positive and negative around singularities. It is possible to have a small or even zero wavelet coefficient when a wavelet overlaps a singularity [92]. This complicates singularity extraction and features modeling. Indeed, the characteristic is inconsistent with some energy-based models which only concern about the magnitudes of coefficients and neglect their sign information.

- Aliasing: The wavelet coefficients are computed via iterated half downsampling operations with filters, which results in substantial aliasing [92]. For face recognition, features are extracted from the wavelet coefficients in wavelet domain. The aliasing makes the extracted features less representative for discrimination.

- Lack of directionality: The standard tensor product construction of 2-D wavelets produces a checkerboard pattern that is simultaneously oriented along several directions [92, 93]. This lack of directional selectivity greatly complicates selection of geometric features like ridges and edges in facial images.

Generally, for achieving translation invariant, it should contain some redundant information in the representing features. This constitutes a trade-off between the amount of possible invariance and the sparseness of the standard wavelet representation. So a robust standard wavelet feature extractor should select a best discriminative feature group with appropriate redundancy. However, searching such a standard wavelet feature extractor is a difficult task.

Furthermore, it should be noted that the primary bottleneck for extracting more powerful features in wavelet domain is the four intrinsic shortcomings of the standard DWT. The solution is to construct more advanced transform which can overcome the shortcomings and generate more discriminant-power features for face recognition. Kingsbury and Selesnick showed that the dual-tree $\mathbb{C}$WT is a good solution to overcome the shortcomings of DWT [92, 94–96]. So we extended the use of the dual-tree $\mathbb{C}$WT to face recognition recently [50].

6.2   Feature extraction based on the dual-tree $\mathbb{C}$WT

There are few studies which adopt the dual-tree $\mathbb{C}$WT technique for face recognition, since it is a new wavelet transform recently studied. The dual-tree $\mathbb{C}$WT has been found to be particularly suitable for image decomposition and representation [92] and used in some fields of image/vedio processing, such as image/vedio denoising [93, 97–99], image fusion [100, 101], handwritten recognition [102], image segmentation [103]. The 2-D dual-tree $\mathbb{C}$WT has the attractive properties for image representation [92, 104], including:

- Approximately analytic and approximately magnitude/phase shift invariance.

- Having smooth, nonnegative magnitudes of complex coefficients without oscillation. This means that the squared magnitudes of coefficients provide an accurate measure of spectral energy, which is consistent to energy-base models.

- Substantially reduced aliasing. The dual-tree (real and imaginary) structure can diminish the effect of the decimation at each scale.

- Good selectivity and directionality in 2-D (also for higher dimensionality), which also render it nearly rotation invariant.

We first make a simple review on the the dual-tree $\mathbb{C}$WT, then briefly introduce our proposed complex-WT-face. Detail researches are shown in [50, 92, 94–96].

6.2.1   The dual-tree $\mathbb{C}$WT

We first consider 1-D case. The 1-D dual-tree $\mathbb{C}$WT is implemented using two critically-sampled real DWTs in parallel on the same data

S, one gives the real part of the transform while the other gives the imaginary part. Figure 6.1 gives a illustration with different sets of filters at different analysis stages.

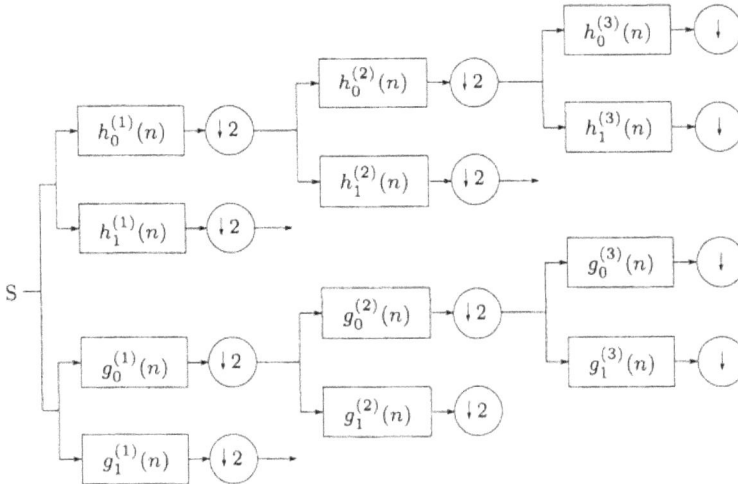

Figure 6.1   Analysis FB for the dual-tree CWT with different sets of filters at different analysis stages [92].

In Figure 6.1, the two real DWTs use two different sets of filters: $h_0^i(n)$, $h_1^i(n)$ denote the low-pass/high-pass filter pair for the upper filter-Bank (FB), and $g_0^i(n)$, $g_1^i(n)$ denote the low-pass/high-pass filter pair for the lower FB at analysis stage i. For simplicity, researchers usually use same filters in different stages, so we cancel the superscript i of $h_0^i(n), h_1^i(n), g_0^i(n), g_1^i(n)$ for convenience. The two sets of filters should satisfy the five conditions: perfect-reconstruction (PR), finite support, vanishing moments, linear-phase, *approximate half-sample delay*. The last one can be represented as $g_0(n) \approx h_0(n - 0.5)$ in mathematics, and considered as the most remarkable property of the dual-tree CWT. Suppose $\psi_h(t)$ and $\psi_g(t)$ are two real wavelets associated with the upper and lower FBs. The approximate half-sample delay property ensures that the complex wavelet $\psi(t) \triangleq \psi_h(t) + j\psi_g(t)$ ($j^2 = -1$) is approximately analytic [105], which brings approximately magnitude/phase shift invariance.

Similar to the 2-D separable DWT, the 2-D dual-tree wavelets associates with the row-column implementation of the 1-D dual-tree wavelets. Although the separable 2-D wavelet transform can represent point-singularities efficiently, it is inefficient for line- and curve-singularities (edges), so it does not possess the optimal properties for facial images. However, the 2-D dual-tree wavelet is not only approximately analytic, but also has good selectivity and directionality, which also render it nearly rotation invariant, thus it is natural for analyzing and processing oriented singularities like edges in facial image. To understand why the dual-tree $\mathbb{C}$WT outperforms the separable wavelet transform, we compare the wavelets associated with the two transforms.

It is known that the separable 2-D wavelet transform is characterized by three wavelets (see (4.3)): $\psi^V(x,y)$ oriented vertically, $\psi^H(x,y)$ oriented horizontally, $\psi^D(x,y)$ oriented diagonally, as shown in Figure 6.2.

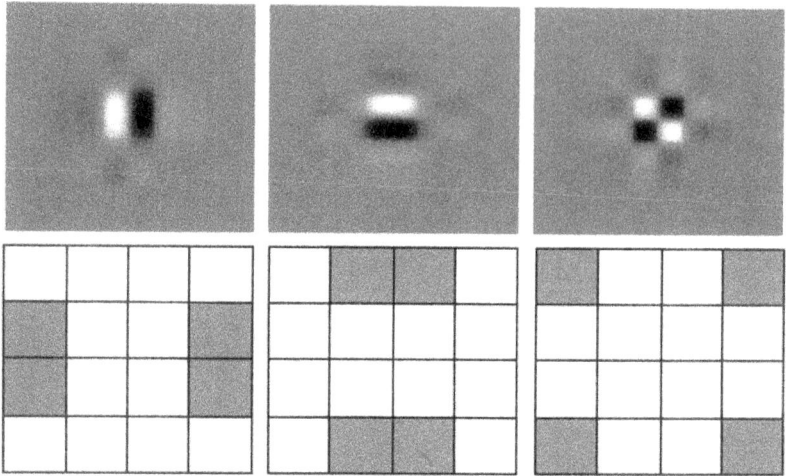

Figure 6.2   Typical wavelets associated with the 2-D separable DWT. The top row illustrates the wavelets in the spatial domain (LH, HL, HH), the second row illustrates the (idealized) support of the spectrum of each wavelet in the 2-D frequency plane. The checkerboard artifact of the third wavelet is apparent in the frequency domain as well as the spatial domain [93].

It is shown that the $\psi^D(x,y)$ wavelet has a checkerboard appearance—it mixes the 45 and −45 degree orientations, that is, the separable DWT fails to isolate these orientations. Because 1-D $\psi(x), \psi(y)$ are real functions whose spectrum must be two-sided, the checkerboard artifact is unavoidable in the frequency domain. Likewise, the checkerboard artifact arises in the spatial domain as well [92, 93].

On the other hand, the 2-D dual-tree $\mathbb{C}$WT is characterized by six wavelets: $\psi_1(x,y) = \phi(x)\psi(y)$, $\psi_2(x,y) = \psi(x)\psi(y)$, $\psi_3(x,y) = \psi(x)\phi(y)$ and their half-conjugate partners $\psi_4(x,y) = \psi(x)\phi(y)$, $\psi_5(x,y) = \psi(x)\overline{\psi(y)}$, $\psi_6(x,y) = \phi(x)\overline{\psi(y)}$, where $\psi(x) \overset{\Delta}{=} \psi_h(x) + j\psi_g(x)$ and $\phi(x) \overset{\Delta}{=} \phi_h(x) + j\phi_g(x)$ are the 1-D complex wavelet function and scale function respectively. All the wavelets $\psi_k(x,y)$ $(1 \leq k \leq 6)$ give six bandpass subbands of complex coefficients at each level, which are strongly oriented at angles of $-75°$, $-45°$, $-15°$, $15°$, $45°$, $75°$ respectively and no checkerboard effect appears, as shown in Figure 6.3.

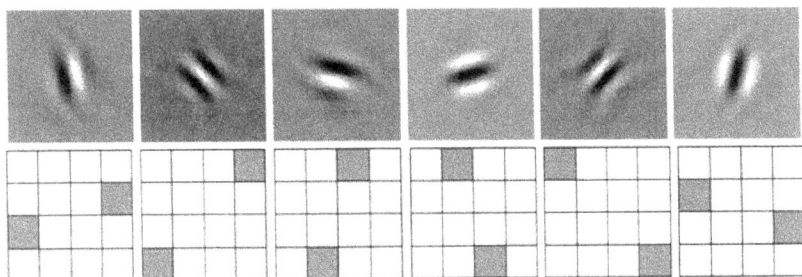

Figure 6.3   Typical wavelets associated with the 2-D dual-tree oriented wavelet transform. The top row illustrates the wavelets in the spatial domain, which are oriented at $-75°$, $-45°$, $-15°$, $15°$, $45°$, $75°$ respectively. The second row illustrates the (idealized) support of the spectrum of each wavelet in the 2-D frequency plane. The absence of the checkerboard phenomenon is observed in both the spatial and frequency domains [93].

It is shown that the 2-D dual-tree $\mathbb{C}$WT succeeds in isolating different orientations, and covers more distinct orientations than does the 2-D separable DWT. Now we consider the expression of $\psi_2(x,y)$ that

$$\begin{aligned} \psi_2(x,y) &= [\psi_h(x) + j\psi_g(x)][\psi_h(y) + j\psi_g(y)] \\ &= \psi_h(x)\psi_h(y) - \psi_g(x)\psi_g(y) + j[\psi_g(x)\psi_h(y) + \psi_h(x)\psi_g(y)]. \end{aligned}$$
(6.1)

(6.1) shows that the real/imaginary part of $\psi_2(x,y)$ can be implemented by subtraction/addition of two 2-D separable DWT in parallel, and other 2-D complex wavelets likewise. So the 2-D dual-tree $\mathbb{C}$WT also has following properties:

- $4\times$redundancy, independent of the number of scales.
- Efficient computation which is only 4 times the standard 2-D separable DWT.

6.2.2   Complex-WT-face

Recently, we have extended the dual-tree $\mathbb{C}$WT to face recognition to compensate for illumination, occlusion, pose and expression variations [50]. Combing the characteristic of the dual-tree $\mathbb{C}$WT and facial images, we proposed an effective feature construction method to provide more representative complex-WT-face features. First, after the 3-level dual-tree $\mathbb{C}$WT, we computed the magnitudes of complex coefficients in all subbands at the third level. Subsequently we introduced a clip method which is based on some statistical information to clip the magnitude features, in order to suppress the effect of abrupt-mutation phenomenon on original complex coefficients. An example of the clipped magnitude subimages of all subbands in the third level is shown in Figure 6.4.

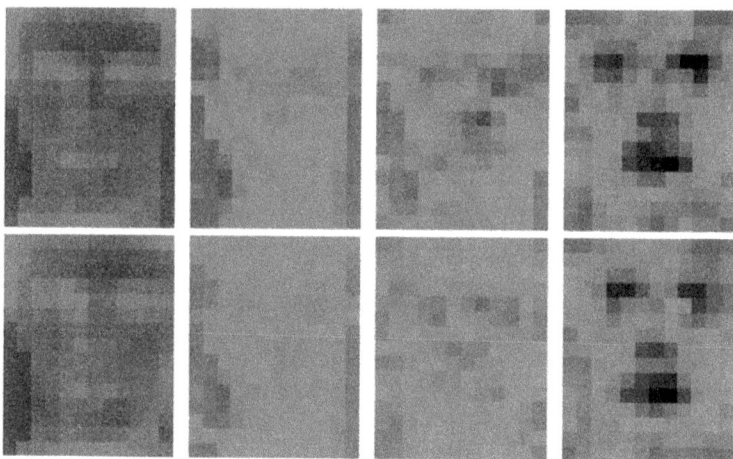

Figure 6.4   An example of the clipped magnitude subimages of all subbands in the third level. In each subimage, the darker means the larger value. The first column is low-frequency pairs; The latter columns are three high-frequency pairs corresponding to three dual-tree wavelet pairs which are oriented at $\mp75°$, $\mp45°$, $\mp15°$.

Then we vectorized the eight magnitude subimages into a large vector, referred to as complex-WT-face. Finally, we use the two-step discriminant analysis to filter and extract enhanced discriminability features. The first one is the componentwise filter which uses a energy-density-based model to select the better discriminant features. The second is the complete linear discriminant analysis (CLDA) which completely extracts both regular and irregular features. Experimental results in [50] show that the proposed discriminant complex-WT-face system can generate enhanced-discriminating-power features, and significantly outperform some other face recognition systems, including fisherFace [21], PCA+CLDA [89], UDA(Ye) [24], ODA(Ye) [24], WaveletFace [51] and LDC [89].

7   Conclusion and discussion

Image-based face recognition is still a very challenging topic after decades of exploration, mostly due to the sensitivity to variations in pose, expressions and different lighting conditions. This survey first provides a brief overview of human face recognition and the structure of pattern recognition system. The analysis of the system and versatility of the wavelet motivate that the wavelet transform can be a powerful tool on two most important components of pattern recognition system: preprocessing and feature extraction.

Whereafter, this survey provides an application-oriented overview of the mathematical foundations of wavelet theory and gives a comprehensive survey of wavelet applications in denoising and feature extraction for face recognition. The object of this survey is to increase familiarity with basic wavelet applications in face recognition and to provide reference sources and examples where the wavelets may be usefully applied to researchers working in the field. Wavelet techniques have a lot of advantages. The representation's sparseness as well as its unique combination of frequency, spatial and scale information recommend it as a technology for facial image localization and recognition. Its multiresolution property is another exciting aspect for applications. More works need to be done on how to best utilize its multiresolution attributes in face recognition. This includes novel extraction/selection of multresolution features and development of more efficient search methods based on the multiresolution [3].

It should also be mentioned that most of works in this survey are based on standard orthonormal wavelet basis. However, we have argued that orthonormal basis may not be the best representation for facial image even though the vanishing moments can help them achieve denoising and dimensionality reduction purpose. Although, orthogonality is the most economical representation without any redundancy, it brings lack of shift-invariance which is very important for recognition. To represent redundant or shift-invariant information, it might be preferable to use appropriate redundant wavelet representations, such as the dual-tree CWT, framelet.

The dual-tree CWT is a valuable enhancement of the traditional real wavelet transform. It has very good attributes, especially approximate shift-invariance and, in higher dimensions, good selectivity and directionality. Moreover, the dual-tree CWT benefits from the vast theoretical, practical, and computational resources that have been developed for the standard DWT. So we forecast that the magnitude and phase of CWT coefficients will be widely exploited to develop new effective wavelet-based algorithms for face recognition.

Another technique should be paid attention to is framelet. In contrast to wavelets, it can be designed with symmetry or anti-symmetry, compact supports and better smoothness. It gives up the orthogonality, but possesses appropriate redundancy. When we decompose data with a framelet, the approach can be viewed as "oversampled" with respect to the Nyquist density in time-frequency space. The redundancy ensures that the frame expansions are more robust to additive noise and quantization degradations and can capture significant data characteristics more efficiently [106]. So we expect its application on face recognition in future researches.

References

[1] J. G. Daugmann, "Two-dimensional spectral analysis of cortical receptive field profile," *Vision Research*, vol. 20, pp. 847–856, 1980.

[2] S. G. Whittaker and J. B. Siegfried, "Origin of wavelets in the visual evoked potential," *Electroencephalogr. Clin. Neurophysiol.*, vol. 55, pp. 91–101, 1983.

[3] R. R. Brooks, L. Grewe and S. S. Iyengar, "Recognition in the wavelet domain: A survey," *Journal of Electronic Imaging*, vol. 10, no. 3, pp. 757–784, Jul. 2001.

[4] J. Wang, G. Naghdy and P. Ogunbona, "Wavelet-based feature-adaptive adaptive resonance theory neural network for texture identification," *Journal of Electronic Imaging*, vol. 6, no. 3, pp. 329–336, 1997.

[5] B. S. Manjunath, "Gabor wavelet transform and application to problems in early vision," *in Conference Record of the Twenty-sixth Asilomar Conference in Signals, Systems and Computers*, vol. 2, pp. 796–800, Oct. 1992.

[6] R. Chellappa, C. L. Wilson and S. Sirohey, "Human and machine recognition of faces: A survey," *Proceedings of the IEEE*, vol. 83, no. 5, pp. 705–740, 1995.

[7] H. Wechsler, P. Phillips, V. Bruce, F. Soulie and T. Huang, *Face Recognition: From Theory to Applications*. Springer-Verlag, 1996.

[8] W. Zhao, R. Chellappa, P. Phillips and A. Rosenfeld, "Face recognition: A literature survey," *ACM Computing Surveys*, vol. 35, no. 4, pp. 399–459, 2003.

[9] X. G. Lu, "Image analysis for face recognition," May 2003, personal notes.

[10] R. Hietmeyer, "Biometric identification promises fast and secure processing of airline passengers," *The Intl Civil Aviation Organization Journal*, vol. 55, no. 9, pp. 10–11, 2000.

[11] A. K. Jain, A. Ross and S. Prabhakar, "An introduction to biometric recognition," *IEEE transactions on circuits and systems for video technology*, vol. 14, no. 1, pp. 4–20, Jan. 2004.

[12] I. Maghiros, Y. Punie, S. Delaitre and et al, *Biometrics at the Frontiers: Assessing the impact on society*. European Commission, Joint Research Center and Institue, 2005.

[13] P. J. Phillips, H. Moon, S. A. Rizvi and P. J. Rauss, "The feret evaluation methodology for face-recognition algorithms," *IEEE Transactions on Pattern Analysis and Machine Intelligence*, vol. 20, no. 10, pp. 1090–1104, 2000.

[14] S. Z. Li and A. K. Jain, "Introduction," in *Handbook of face recognition*, S. Z. Li and A. K. Jain, Eds. Springer-Verlag, 2005.

[15] Y. Moses, Y. Adini and S. Ullman, "Face recognition: The problem of compensating for changes in illumination direction," in *Proceedings of the European Conference on Computer Vision*, vol. A, pp. 286–296, 1994.

[16] I. S. Bruner and R. Tagiuri, "The perception of people," in *Handbook of Social Psychology*, G. Lindzey, Ed. Addison-Wesley, vol. 2, pp. 634–654, 1954.

[17] W. W. Bledsoe, "The model method in facial recognition," Panoramic Research Inc., Palo Alto, CA, Technical Report 15, 1964.

[18] M. D. Kelly, "Visual identification of people by computer," Stanford AI Project, Stanford, CA, Technical Report AI-130, 1970.

[19] T. Kanade, "Computer recognition of human faces," *Interdisciplinary Systems Research*, vol. 47, 1977.

[20] M. Turk and A. Pentland, "Eigenfaces for recognition," *Journal of Cognitive Neuroscience*, vol. 3, pp. 72–86, 1991.

[21] P. N. Belhumeur, J. P. Hespanha and D. J. Kriegman, "Eigenfaces versus fisherfaces: recognition using class specific linear projection," *IEEE Transactions on Pattern Analysis and Machine Intelligence*, vol. 19, no. 7, pp. 711–720, Jul. 1997.

[22] J. Yang and J. Y. Yang, "Why can lda be performed in pca transformed space?" *Pattern Recognition*, vol. 36, no. 2, pp. 563–566, 2003.

[23] A. M. Martinez and A. C. Kak, "Pca versus lda," *IEEE Transaction on Pattern Analysis and Machine Intelligence*, vol. 23, no. 2, pp. 228–233, 2001.

[24] J. P. Ye, "Characterization of a family of algorithms for generalized discriminant analysis on undersampled problems," *Journal of Machine Learning Research*, vol. 6, pp. 483–502, 2005.

[25] D. Q. Dai and P. C. Yuen, "Regularized discriminant analysis and its applications to face recognition," *Pattern Recognition*, vol. 36, no. 3, pp. 845–847, 2003.

[26] A. Pentland, B. Moghaddam and T. Starner, "View-based and modular eigenspaces for face recognition," in *Proceedings of IEEE Conference on Computer Vision and Pattern Recognition*, 1994.

[27] P. Penev and J. Atick, "Local feature analysis: A general statistical theory for objecct representation," *Network: Computation in Neural Systems*, vol. 7, no. 3, pp. 477–500, 1996.

[28] A. Lanitis, C. J. Taylor and T. F. Cootes, "Automatic face identification system using flexible appearance models," *Image and Vision Computing*, vol. 13, no. 5, pp. 393–401, 1995.

[29] G. Toussaint, "Introduction to pattern recognition," http://cgm.cs.mcgill.ca/godfried/teaching/pr-web.html.

[30] A. R. Webb, *Statistical Pattern Recognition*, 2nd ed. John Wiley & Sons, Ltd., 2002.

[31] I. Daubechies, *Ten Lectures on Wavelets*. New York: Society for Industrial and Applied Mathematics, 1992.

[32] F. Abramovich, T. Bailey and T. Sapatinas, "Wavelet analysis and its statistical applications," *Journal of the Royal Statistical Society*, vol. 49, no. 1, pp. 1–29, Mar. 2000.

[33] S. Mallat, *A Wavelet Tour of Signal Processing*, 2nd ed. San Diego: Academic Press, 1999.

[34] R. R. Coifman and Y. Meyer, "Orthonormal wavelet packet bases," 1990, preprint.

[35] M. Vetterli and J. Kovaèevi, *Wavelets and Subband coding*. Prentice Hall, 1995.

[36] H. Choi and R. G. Baraniuk, "Wavelet statistical models and besov spaces," in *Nonlinear Estimation and Classification*, D. D. Denison, M. H. Hansen, C. C. Holmes, B. Mallick, and B. Yu, Eds. New York: Springer-Verlag, 2003, pp. 9–29.

[37] D. Q. Dai and H. Yan, "Wavelets and face recognition," in *Face Recognition*, K. Delac and M. Grgic, Eds. I-Tech Education and Publishing, Jun. 2007, pp. 59–74.

[38] T. Li, Q. Li, S. Zhu and M. Ogihara, "A survey on wavelet applications in data mining," *ACM SIGKDD Explorations Newsletter*, vol. 4, no. 2, pp. 49–68, 2002.

[39] J. Koenderink, "The structure of images," in *Biological Cybernrtics*. New York: Springer-Verlag, 1984.

[40] S. G. Mallat, "A theory for multiresolution signal decomposition: the wavelet representation," *IEEE transactions on pattern analysis and machine intelligence*, vol. 11, no. 7, pp. 674–693, 1989.

[41] I. Daubechies, "The wavelet transform, time-frequency localization and signal analysis," *IEEE Transactions on Information Theory*, vol. 36, no. 5, pp. 961–1005, 1990.

[42] S. C. Zhang, C. Q. Zhang and Q. Yang, "Data preparation for data mining," *Applied Artificial Intelligence*, vol. 17, pp. 375–381, 2003.

[43] D. L. Donoho and I. M. Johnstone, "Minimax estimation via wavelet shrinkage," *Annals of Statistics*, vol. 26, no. 3, pp. 879–921, Jun. 1998.

[44] J. L. Starck, E. J. Candes and D. L. Donoho, "The curvelet transform for image denoising," *IEEE Transactions on Image Processing*, vol. 11, no. 6, pp. 670–684, Jun. 2002.

[45] S. G. Chang, B. Yu and M. Vetterli, "Spatially adaptive wavelet thresholding with context modeling for image denoising," *IEEE Transactions on Image Processing*, vol. 9, pp. 1522–1531, 2000.

[46] V. Strela, "Denoising via block wiener filtering in wavelet domain," in *3rd European Congress of Mathematics*. Barcelona: Birkhauser Verlag, Jul. 2000.

[47] D. L. Donoho and I. M. Johnstone, "Ideal spatial adaptation via wavelet shrinkage," *Biometrika*, vol. 81, no. 9, pp. 425–455, 1993.

[48] R. T. Ogden, *Essential Wavelets for Statistical Applications and Data Analysis*. Boston: Birkhaeuser, 1997.

[49] L. L. Shen and L. Bai, "A review on gabor wavelets for face recognition," *Pattern Analysis and Applications*, vol. 9, pp. 273–292, 2006.

[50] C. C. Liu and D. Q. Dai, "Discriminant dual-tree complex wavelet features for face recognition," Dec. 2007, submitted.

[51] J. T. Chien and C. C. Wu, "Discriminant waveletfaces and nearest feature classifiers for face recognition," *IEEE Transactions on Pattern Analysis and Machine Intelligence*, vol. 24, no. 12, pp. 1644–1649, 2002.

[52] C. Garcia, G. Zikos and G. Tziritas, "A wavelet-based framework for face recognition," in *Proc of the Workshop on Advances in Facial Image Analysis and Recognition Technology*. Freiburg Allemagne: 5th European Conference on Computer Vision (ECCV'98), pp. 84–92, 1998.

[53] ——, "Wavelet packet analysis for face recognition," *Image and Vision Computing*, vol. 18, pp. 289–297, 2000.

[54] N. H. Foon, Y. H. Pang, A. T. B. Jin and D. N. C. Ling, "An efficient method for human face recognition using wavelet transform and zernike moments," in *Proceedings of the International Conference on Computer Graphics, Imaging and Visualization (CGIV04)*, pp. 65–69, 2004.

[55] L. M. Cui, Y. Y. Tang, F. C. Liao and X. F. Du, "Face recognition based on wavelet packet decomposition and support vector machines," in *Proceedings of the 2007 International Conference on Wavelet Analysis and Pattern Recognition*, Beijing, Nov. 2007.

[56] Q. X. Xu and J. Wei, "Application of wavelet energy feature in facial expression recognition," *IEEE International Workshop on Anti-counterfeiting, Security, Identification*, pp. 169–174, Apr. 2007.

[57] C. J. Chen and J. S. Zhang, "Wavelet energy entropy as a new feature extractor for face recognition," in *Fourth International Conference on Image and Graphics*, pp. 616–619, Aug. 2007.

[58] S. Kullback and R. A. Leibler, "On information and sufficieny," *Annals of Mathematical Statistics*, vol. 22, pp. 79–86, 1951.

[59] G. C. Feng, P. C. Yuen and D. Q. Dai, "Human face recognition using pca on wavelet subband," *Journal of Electronic Imaging*, vol. 9, no. 2, pp. 226–233, Apr. 2000.

[60] H. Wang, S. Yang and W. Liao, "An improved pca face recognition algorithm based on the discrete wavelet transform and the support vector machines," in *International Conference on Computational Intelligence and Security Workshops*, pp. 308–311, 2007.

[61] W. Puyati, S. Walairacht and A. Walairacht, "Pca in wavelet domain for face recognition," in *The 8th International Conference on Advanced Communication Technology (ICACT)*, vol. 1, pp. 450–455, 2006.

[62] Y. Z. Shen, Y. Chen, H. Feng and T. N. He, "A recognition algorithm of facial image based on wavelet subbands and dicision fusion," in *Proceedings of the Second International Conference on Machine Learning and Cybernetics*, Nov. 2003.

[63] M. Safari, M. T. Harandi and B. N. Araabi, "A svm-based method for face recognition using a wavelet-pac representation of faces," in *International Conference on Image Processing (ICIP)*, pp. 853–856, 2004.

[64] Q. Yin, Z. Y. Yuan, Y. Kong and P. Guo, "Face recognition research based on anti-symmetrical wavelet and eigenface," in *Proceedings of the Sixth International Conference on Machine Learning and Cybernetics*, Hong Kong, pp. 366–371, Aug. 2007.

[65] K. A. Kim, S. Y. Oh and H. C. Choi, "Facial feature extraction using pca and wavelet multi-resolution images," in *Proceedings of the Sixth IEEE International Conference on Automatic Face and Gesture Recognition (FGR'04)*, pp. 439–444, May 2004.

[66] D. Q. Dai and P. C. Yuen, "Wavelet-based 2-parameter regularized discriminant analysis for face recognition," in *AVBPA*, J. Kittler and M. S. Nixon, Eds. Berlin Heidelberg: Springer-Verlag, vol. 2688, pp. 137–144, 2003.

[67] ——, "Wavelet based discriminant analysis for face recognition," *Applied Mathematics and Computation*, vol. 175, pp. 307–318, 2006.

[68] W. G. Cao, K. Jiang, Z. H. Yu and B. Y. Sun, "Human face recognition using generalized kernel fisher discriminant and wavelet transform," in *Proceedings of the 2006 IEEE International Conference on Information Acquisition*, Shandong, China, pp. 1258–1262, Aug. 2006.

[69] H. K. Ekenel and B. Sanker, "Multiresolution face recognition," *Image and Vision Computing*, vol. 23, pp. 469–477, 2005.

[70] N. Shams, I. Hosseini, M. S. Sadri and E. Azarnasab, "Low cost fpga-based highly accurate face recognition system using combined wavelets with subspace methods," in *IEEE International Conference on Image Processing*, pp. 2077–2080, Oct. 2006.

[71] B. L. Zhang, H. H. Zhang and S. S. Ge, "Face recognition by applying wavelet subband representation and kernel associative memory," *IEEE Transactions on Neural Networks*, vol. 15, no. 1, pp. 166–177, 2004.

[72] J. Lin, J. P. Li and M. Ji, "Robust face recognition by wavelet features and model adaptation," in *Proceedings of the 2007 International Conference on Wavelet Analysis and Pattern Recognition*, Beijing, China, pp. 1638–1643, Nov. 2007.

[73] M. R. M. Rizk, O. Said, R. El-Sayed, and K. Sobhy, "Wavelets and neural networks based face recognition system," in *Proceedings of the 46th IEEE*

International Midwest Symposium on Circuits and Systems, vol. 2, pp. 965–968, Dec. 2003.

[74] Z. B. Zhang, S. L. Ma and D. Y. Wu, "The application of neural network and wavelet in human face illumination compensation," *Lecture Notes in Computer Science*, vol. 3497, pp. 828–835, 2005.

[75] V. Krüger and G. Sommer, "Wavelet networks for face processing," *Journal of the Optical Society of America*, vol. 19, no. 6, pp. 1112–1119, 2002.

[76] A. Amira and P. Farrell, "An automatic face recognition system based on wavelet transforms," in *IEEE International Symposium on Circuits and Systems*, vol. 6, pp. 6252–6255, 2005.

[77] M. Sujaritha and S. Annadurai, "Face recognition using wavelet transform and locally discriminating projection," *International Conference on Computational Intelligence and Multimedia Applications*, vol. 2, pp. 436–440, Dec. 2007.

[78] C. Y. Xie, M. Savvides and B. V. K. Vijayakumar, "Quaternion correlation filters for face recognition in wavelet domain," in *IEEE International Conference on Acoustics, Speech, and Signal Processing*, vol. 2, pp. 85–88, Mar. 2005.

[79] C. Xie and B. Vijayakumar, "Quaternion correlation filters for color face recognition," in *Conference on Security, Steganography, and Watermarking of Multimedia Contents*. San Jose, CA: SPIE Symposium on Electronic Imaging, pp. 16–20, Jan 2005.

[80] K. C. Kwak and W. Pedrycz, "Face recognition using fuzzy integral and wavelet decomposition method," *IEEE Transactions on Systems, Man, and Cybernetics–Part B: Cybernetics*, vol. 34, no. 4, pp. 1666–1675, Aug. 2004.

[81] J. H. Lai, P. C. Yuen and G. C. Feng, "Face recognition using holistic fourier invariant features," *Pattern Recognition*, vol. 34, pp. 95–109, 2001.

[82] R. R. Coifman and M. V. Wicherhauser, "Entropy-based algorithm for best basis selection," *IEEE Transactions on Information Theory*, vol. 38, no. 2, pp. 713–718, Mar. 1992.

[83] S. Mallat and Z. Zhang, "Matching pursuit in a time–frequency dictionary," *IEEE Transactions on Signal Processing*, vol. 41, pp. 3397–3415, 1993.

[84] N. Saito and R. R. Coifman, "Local discriminant bases," *Proceedings of SPIE*, vol. 2303, pp. 2–14, 1994.

[85] ——, "Local discriminant bases and their applications," *Journal of Mathematical Imaging and Vision*, vol. 5, no. 4, pp. 337–358, 1995.

[86] N. Saito, R. R. Coifman, F. B. Geshwind and F. Warner, "Discriminant feature extraction using empirical probability density estimation and a local basis library," *Pattern Recognition*, vol. 35, pp. 2841–2852, 2002.

[87] A. Z. Kouzani, F. He and K. Sammut, "Wavelet packet face representation and recognition," in *IEEE International Conference on Systems, Man, and Cybernetics*, vol. 2, pp. 1614–1619, Oct. 1997.

[88] R. Bhagavatula and M. Savvides, "Pca vs. automatically pruned wavelet-packet pca for illumination tolerant face recognition," in *Fourth IEEE Workshop on Automatic Identification Advanced Technologies*, pp. 69–74, 2005.

[89] C. C. Liu, D. Q. Dai and H. Yan, "Local discriminant wavelet packet coordinates for face recognition," *Journal of Machine Learning Research*, vol. 8, pp. 1165–1195, 2007.

[90] C. Nastar and N. Ayach, "Frequency-based nonrigid motion analysis," *IEEE Transactions on Pattern Analysis and Machine Intelligence*, vol. 18, pp. 1067–1079, 1996.

[91] S. Mallat, "Wavelets for a vision," *Proceedings of The IEEE*, vol. 84, no. 4, pp. 604–614, 1996.

[92] I. W. Selesnick, R. G. Baraniuk and N. Kingsburg, "The dual-tree complex wavelet transform – a coherent framework for multiscale signal and image processing," *IEEE Signal Processing Magazine*, vol. 22, no. 6, pp. 123–151, Nov. 2005.

[93] I. W. Selesnick and K. Y. Li, "Video denoising using 2d and 3d dual-tree complex wavelet transforms," in *Wavelets: Applications in Signal and Image Processing X*, M. A. Unser, A. Aldroubi, and A. F. Laine, Eds., vol. 5207. San Diego: Proceedings of SPIE, Aug. 2003.

[94] N. G. Kingsbury, "The dual-tree complex wavelet transform: A new technique for shift invariance and directional filters," in *Proceedings 8th IEEE DSP Workshop*, no. 86, Utah, Aug. 1998.

[95] ——, "Complex wavelets for shift invariant analysis and filtering of signals," *Applied Computational Harmonic Analysis*, vol. 10, no. 3, pp. 234–253, May 2001.

[96] I. W. Selesnick, "The design of approximate hilbert transform pairs of wavelet bases," *IEEE Transactions on Signal Processing*, vol. 50, no. 5, pp. 1144–1152, May 2002.

[97] F. X. Yan, S. L. Peng and L. Z. Cheng, "Dual-tree complex wavelet hidden markov tree model for image denoising," *Electronics Letters*, vol. 43, no. 18, pp. 973–975, Aug. 2007.

[98] B. Chen, Z. X. Geng, Y. Yang and T. S. Shen, "Dual-tree complex wavelets transforms for image denoising," in *Eighth ACIS International Conference on Software Engineering, Artificial Intelligence, Networking, and Parallel/Distributed Computing*, pp. 70–74, 2007.

[99] H. Rabbani and M. Vafadust, "Image/video denoising based on a mixture of laplace distributions with local parameters in multidimensional complex wavelet domain," *Signal Processing*, vol. 88, pp. 158–173, 2008.

[100] J. J. Lewis, R. J. O. Callaghan, S. G. Nikolov, D. R. Bull and N. Canagarajah, "Pixel- and region-based image fusion with complex wavelets," *Information Fusion*, vol. 8, pp. 119–130, 2007.

[101] S. Ioannidou and V. Karathanassi, "Investigation of the dual-tree complex and shift-invariant discrete wavelet transforms on quickbird image fusion," *IEEE geoscience and remote sensing letters*, vol. 4, no. 1, pp. 166–170, Jan. 2007.

[102] G. Y. Chen and W. F. Xie, "Pattern recognition with svm and dual-tree complex wavelets," *Image and Vision Computing*, vol. 25, pp. 960–966, 2007.

[103] E. H. S. Lo, M. R. Pickering, M. R. Frater and J. F. Arnold, "Image segmentation using invariant texture features from the double dyadic dual-tree complex wavelet transform," in *IEEE International Conference on Acoustics, Speech and Signal Processing*, vol. 1, pp. I–609–I–612, Apr. 2007.

[104] N. G. Kingsbury, "The dual-tree complex wavelet transform: a new efficient tool for image restoration and enhancement," in *Proceedings of European Signal Processing Conference*, Rhodes, pp. 319–322, 1998.

[105] I. W. Selesnick, "Hilbert transform pairs of wavelet bases," *IEEE Signal Processing Letters*, vol. 8, no. 6, pp. 170–173, Jun. 2001.

[106] V. K. Goyal, J. Kovaèevi and J. A. Kelner, "Quantized frame expansions with erasures," *Journal of Applied and Computational Harmonic Analysis*, vol. 10, no. 3, pp. 203–233, 2001.

Hilbert-Huang Transform: Its Background, Algorithms and Applications*

Lihua Yang

School of Mathematics and Computing Science
Sun Yat-Sen (Zhongshan) University, China
Email: mcsylh@mail.sysu.edu.cn

Abstract

This article is based on the lecture for graduate students at the ISFMA Symposium on Wavelet Methods in Mathematical Analysis and Engineering at Zhuhai campus of Sun Yat-Sen (Zhongshan) University, China, in August 2007. It aims to give an intuitive introduction of the background of the Hilbert-Huang transform and a summary of our recent works on the relevant theoretic questions and applications to pattern recognition.

1 Background: amplitude, phase and frequency

1.1 Signal, period and frequency

A signal is a function of time t. For a signal $s(t)$, if there exists a positive number T such that

$$s(t + T) = s(t) \quad (\forall t \in \mathbb{R}),$$

where $\mathbb{R}$ denotes the set of all the real numbers, then it is called a periodic signal and T is its period. Let T be the smallest positive period of $s(t)$. Then the frequency and angular frequency of $s(t)$ are defined as

$$f := \frac{1}{T} \quad \text{and} \quad \omega := \frac{2\pi}{T}$$

respectively.

The frequency f represents the velocity of oscillatory of $s(t)$. If the unit of time is second, then the unit of frequency is hertz (namely oscillatory times/second), denoted by Hz. Intuitively, frequency characterizes

---

*This work is supported in part by NSFC (Nos. 10631080, 60475042).

the oscillatory velocity of a signal. Nobody doubts the reasonableness of
the frequency defined above for a periodic signal. Figure 1.1 illustrates
two harmonic signals of different frequencies. Figure 1.1(a) is the signal
$s(t) = \cos(2\pi t)$, whose smallest positive period is $T = 1$ and frequency
is $f = \frac{1}{T} = 1$; Figure 1.1(b) is the signal $s(t) = \cos(4\pi t)$, whose smallest
positive period is $T = 1/2$ and frequency is $f = \frac{1}{T} = 2$.

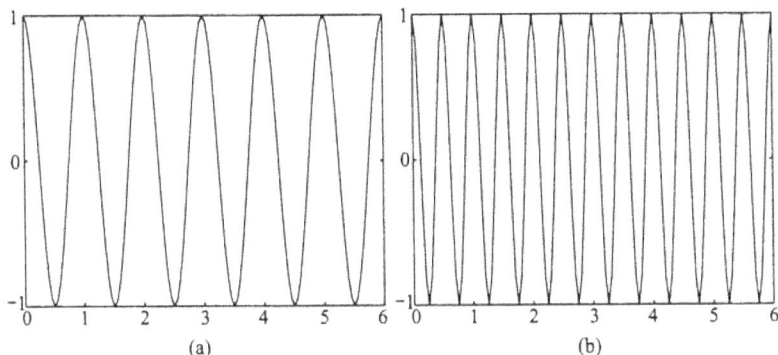

Figure 1.1   The harmonic signal $s(t) = \cos(\frac{2\pi}{T}t)$. (a) $T = 1$, $f = 1$; (b) $T = 1/2$, $f = 2$.

Figure 1.2 also shows two periodic signals having different frequencies, which are piecewise linear functions. The smallest positive period
in Figure 1.2(a) is $T = 1$ and the corresponding frequency is $f = \frac{1}{T} = 1$;
while the smallest positive period in Figure 1.2(b) is $T = 1/2$, and the
corresponding frequency is $f = \frac{1}{T} = 2$. Thus the concept of frequency
can characterize the velocity of the oscillation of a signal but cannot tell
us how it oscillates within a peroid.

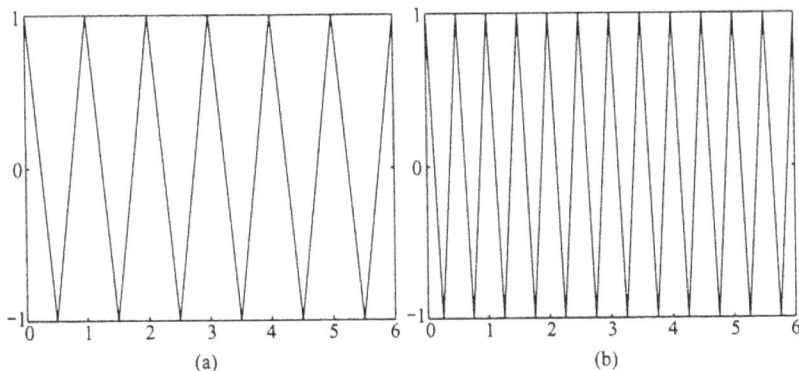

Figure 1.2   Piecewise line periodic oscillations. (a) $T = 1$, $f = 1$; (b) $T = 1/2$, $f = 2$.

Periodic harmonic waves are the most important signals in practice, such as magnetic waves, sound waves, etc. The mathematical foundation of harmonic waves are the well-known Fourier analysis. It is proved that for any $s \in L_T^2$ (the space of all the T-periodic and finite energy signals), there holds

$$s(t) = \sum_{k \in \mathbb{Z}} c_k e^{ik\frac{2\pi}{T}t}, \qquad c_k := \frac{1}{T} \int_0^T s(t) e^{-ik\frac{2\pi}{T}t} dt,$$

where $\mathbb{Z}$ is the set of all the integers.

1.2   Signal transmission and modulation

Physical signals are transmitted by electromagnetic waves (Figure 1.3). They must be modulated before transmission since (1) most natural signals are usually of lower frequencies (the frequencies of human speech are generally ranges from 300Hz $\sim$ 3000Hz). To received such signals, antennas must be 10 $\sim$ 100 kilometers long (L > wavelength/10) if no modulation. (2) To avoid interference among signals with similar frequencies, and (3) with modulation, orthogonal frequency division multiplexing technology (OFDM) can be used to save the physical channels.

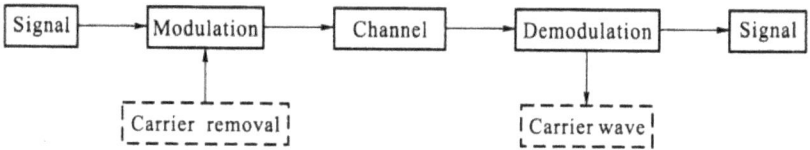

Figure 1.3   Signal transmission.

There are usually three kinds of modulation for signals. Let $m(t)$ be the signal to be transmitted and $s(t) = m_0 \cos \omega_c t$ be a carrier wave. Then $m(t)$ can be modulated as:

- Amplitude modulation: $s_{AM}(t) = [m_0 + m(t)] \cos(\omega_c t)$;
- Frequency modulation: $s_{FM}(t) = m_0 \cos[\omega_c t + \int_{-\infty}^t K_{FM} m(\tau) d\tau]$;
- Phase modulation: $s_{PM}(t) = m_0 \cos[\omega_c t + m(t)]$.

For the receiver, to understand, analyze and process a signal, demodulation is necessary. That means, for a given signal $s(t)$ one needs to find the instantaneous amplitude $\rho(t)$ and instantaneous phases $\theta(t)$ such that $s(t) = \rho(t) \cos \theta(t)$. The derivative of $\theta(t)$ is called the instantaneous frequency: $\omega(t) = \theta'(t)$.

For $s(t) = A \cos(\omega t)$, it is obviously seen that $\rho(t) \equiv A$ and $\omega(t) = (\omega t)' = \omega$, which means that the amplitude and frequency of $s(t)$ are respectively A and ω.

Let us consider the following signal:

$$s(t) = \sqrt{t}(2 + \sin t)\cos(5t).$$

According to the expression we have $\rho(t) = \sqrt{t}(2 + \sin t)$ (the above line in Figure 1.4) and $\omega(t) = \theta'(t) = 5$.

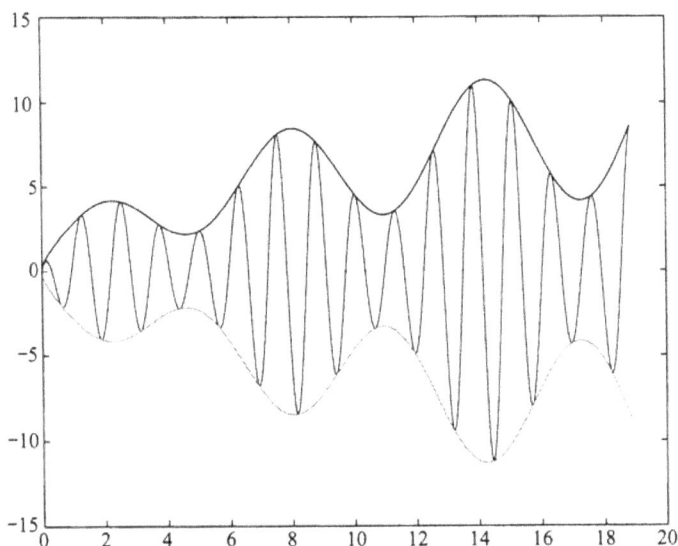

Figure 1.4   A signal whose frequency and amplitude are time-varing.

We turn to consider another example shown in the left of Figure 1.5. Our question is: What is the instantaneous frequency and amplitude of the signal? Since

$$s(t) = m(t)\cos(t) \quad \text{with} \quad m(t) = \cos^2(2t),$$

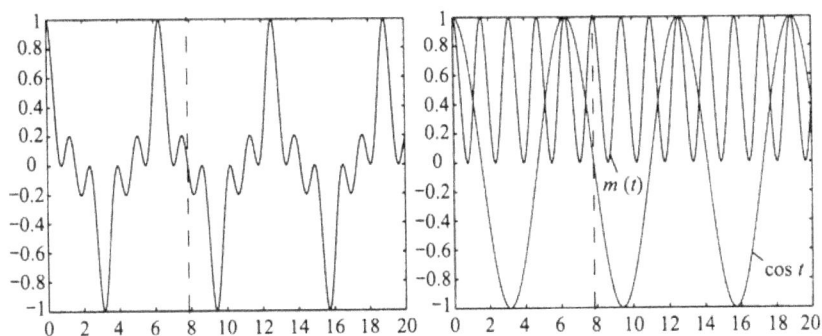

Figure 1.5   Left: original signal; Right: $m(t)$ and $\cos t$.

we have $\rho(t) = m(t) = \cos^2(2t)$ (the red line) and $\omega(t) = 1$. It gives a fantastic demodulation (contradicts the physical observation).

In general, for a given real-valued signal $s(t)$, there are infinite pairs $(\rho(t), \theta(t))$ satisfying $s(t) = a(t) \cos \theta(t)$. It is difficult to define the physically meaningful instantaneous frequency precisely in mathematics.

1.3   The existence of instantaneous frequency for aperiodic signals

Most signals we encounter in practice are aperiodic. These signals usually contain plentiful oscillation. Can we define frequency for such signals? Physically, are there frequencies for aperiodic signals? Figure 1.6 shows the waveforms of the signal $s(t) = \cos(2\pi(t+1)t)$. It is easy to see that this signal oscillates forever, faster and faster as the time increases. Therefore, in physics, there must exists a time-varying *frequency* for this signal.

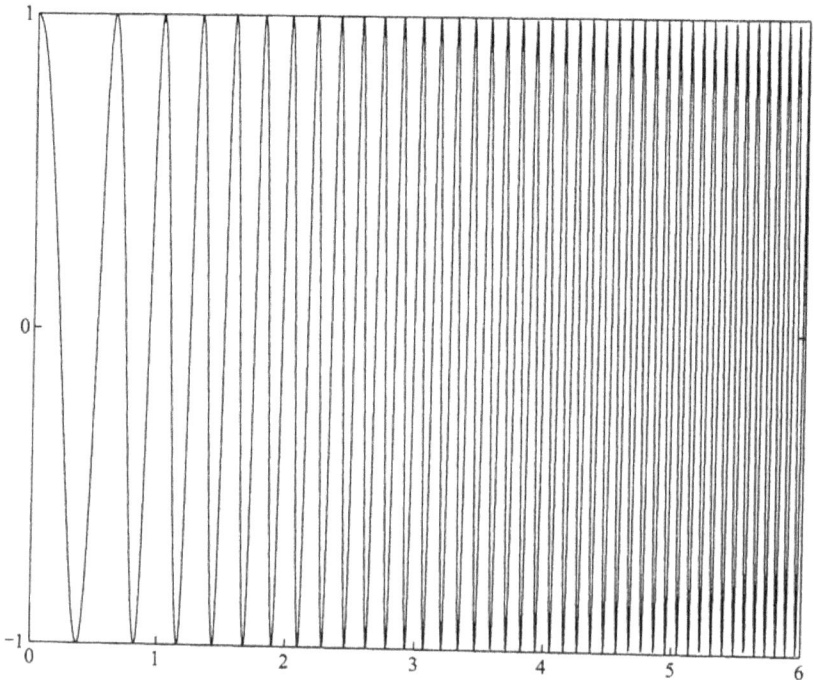

Figure 1.6   A example of time-varying physical frequency.

Our question is: How to define the time-varying frequency (or instantaneous frequency) in mathematics?

A complex-valued signal is called a complex signal for simplicity. It is easy to define the instantaneous frequency of a complex signal. The simplest complex signal is the harmonic signal $s(t) = Ae^{i(\omega t+\theta_0)}$, where $A > 0$ and $\omega > 0$. It is a typical periodic signal, whose amplitude is A, period is $T = 2\pi/\omega$, frequency is $f := \omega/2\pi$, and angular frequency is $\nu := 2\pi/T = \omega$. The graph of the signal can be seen as the orbit a particle rotates anticlockwise around the unit circle. At the instant t, the particle is at the location whose angle is $\theta(t) := \omega t + \theta_0$, so we call $\omega t + \theta_0$ the phase of $s(t)$. When $t = 0$, the phase is called the initial phase. The particle rotates uniformly and anticlockwise with the angle velocity ω.

A general complex signal can be written as $s(t) = A(t)e^{i\theta(t)}$, where $A(t) = |s(t)| \geq 0$, $\theta(t) \in \mathbb{R}$. The graph of the signal can still be understood with the circular motion model of a particle discussed above. Other than the harmonic signal, its amplitude and phase vary with time. The track of the particle is not a circle and the motion is not uniform anymore. The instantaneous angle velocity is $\omega(t) := \theta'(t)$. If $\theta'(t) > 0$, it moves anticlockwise at the moment; otherwise, it does clockwise.

Definition 1.1. *For a complex signal $s(t) = A(t)e^{i\theta(t)}$ with $A(t) \geq 0$ and $\theta(t) \in \mathbb{R}$, $A(t)$ and $\theta(t)$ are called the instantaneous amplitude and instantaneous phase of $s(t)$ respectively. If $\theta(t)$ is differentiable, $\omega(t) := \theta'(t)$ is called the instantaneous frequency of $s(t)$ at instant t.*

1.4   Complex extension and demodulation by Hilbert transform

A real-valued signal is called a real signal. Signals in reality are usually real signals. A real signal can be seen as a to-and-fro motion of a particle in the real axis. Different from complex signals, it is very difficult to define the instantaneous frequency of a real signal since we do not know what is the phase of a real signal. We can not define the instantaneous frequency directly from a real signal. However, if we can find a complex signal whose real part is the real signal, then it is reasonable to define the the instantaneous frequency of the real signal as that of the complex signal.

It is well known that the instantaneous frequency of the harmonious signal $\cos(\omega t+\theta_0)$ is ω. Its reasonable complex signal should be $e^{i(\omega t+\theta_0)}$. Although we can choose the imaginary part other than $\sin(\omega t + \theta_0)$ to obtain different phase and instantaneous frequency, this one is the most meaningful.

Given a signal $s(t)$, its Fourier transform is defined as

$$\hat{s}(\omega) := \int_{\mathbb{R}} s(t)e^{-i\omega t}dt = \int_{\mathbb{R}} s(t)\overline{e^{i\omega t}}dt = \langle s(t), e^{i\omega t}\rangle. \qquad (1.1)$$

Physically, $\hat{s}(\omega)$ is the projection of $s(t)$ on the complex harmonic signal $e^{i\omega t}$. The signal can be determined by all the Fourier projections

$\{\hat{s}(\omega)|\omega \in \mathbb{R}\}$. In fact, we have the inverse Fourier transform as follows:

$$s(t) = \frac{1}{2\pi} \int_{\mathbb{R}} \hat{s}(\omega)e^{i\omega t} d\omega. \tag{1.2}$$

$\{\hat{s}(\omega)|\omega \in \mathbb{R}\}$ is called the Fourier spectrum of $s(t)$.

It is easy to prove that a real signal $s(t)$ satisfies $\hat{s}(-\omega) = \overline{\hat{s}(\omega)}$. Thus, we have the following conclusion:

$$s(t) = \frac{1}{2\pi} \int_{\mathbb{R}} \hat{s}(\omega)e^{i\omega t} d\omega = \frac{1}{\pi} \mathrm{Re}\left[\int_0^\infty \hat{s}(\omega)e^{i\omega t} d\omega\right].$$

Denote

$$a(t) := \frac{1}{\pi} \int_0^\infty \hat{s}(\omega)e^{i\omega t} d\omega, \tag{1.3}$$

it is proved that $a(t)$ is the unique complex signal satisfying the following conditions:

$$s(t) = \mathrm{Re}\, a(t), \quad \hat{a}(\omega) = \begin{cases} 2\hat{s}(\omega), & \omega \geq 0, \\ 0, & \omega < 0, \end{cases}$$

where $\mathrm{Re}(z)$ standards for the real part of the complex number z.

Definition 1.2 (c.f. [10]). *A complex signal $z(t)$ is called an analytic signal, if* $\mathrm{supp}\hat{z} \subset [0,\infty)$; *Similarly, it is called dual analytic if* $\mathrm{supp}\hat{z} \subset (-\infty, 0]$.

Obviously, the complex signal $a(t)$ defined by Equation (1.3) is an analytic signal.

Theorem 1.3 (c.f. [10]). *A complex signal $z(t) = x(t)+iy(t)$ is analytic if and only if $Hz = -iz$, that is, $y(t) = Hx(t)$, where H is the Hilbert transform defined by*

$$Hf(x) = \frac{1}{\pi}\mathrm{p.v.}\int_{\mathbb{R}} \frac{f(x-t)}{t} dt, \tag{1.4}$$

where p.v. denotes the Cauchy principal value.

It is well known that the Hilbert transform is closed in $L^p(\mathbb{R})$ ($1 < p < \infty$), the space of all the p-power integrable function f defined on $\mathbb{R}$, namely,

$$\|f\|_p := \left(\int_{\mathbb{R}} |f(x)|^p dx\right)^{1/p} < \infty.$$

Definition 1.4. *Given a real signal $s(t)$, the complex signal $a(t) = s(t) + iHs(t)$ is called the analytic signal of $s(t)$, operator A defined by $As := s + iHs$ is called the analytic operator, and the amplitude, phase and the instantaneous frequency of $a(t)$ are defined as the amplitude, phase and the instantaneous frequency of $s(t)$ respectively.*

Theorem 1.5. *All the following signals are analytic: (1) The linear combination of finite analytic signals; (2) The product of two analytic signals; (3) The convolution of an analytic signal with any other signals.*

Example 1.1. *Suppose* $\text{Im} z < 0$, $k \in \mathbb{N}$, *then* $s(t) := (t - z)^{-k}$ *is an analytic signal.*

Example 1.2. *Suppose* $z_1, \cdots, z_N$ *are distinct points in the complex plane satisfying* $\text{Im} z_k > 0$ *and* $m_k \in \mathbb{N}$ ($k = 1, 2, \cdots, N$). *Then the Blaschke product defined by*

$$b(t) := \prod_{k=1}^{N} \left(\frac{t - z_k}{t - \overline{z_k}} \right)^{m_k}$$

is an analytic signal.

The analytic signal provides us a tool to define the instantaneous amplitude, phase and frequency of a real signal. Below are some examples.

Example 1.3. *Let* $x(t) = \cos(t) + 2\cos(1.2t)$. *Its instantaneous amplitude is shown in Figure 1.7, which is physically meaningful.*

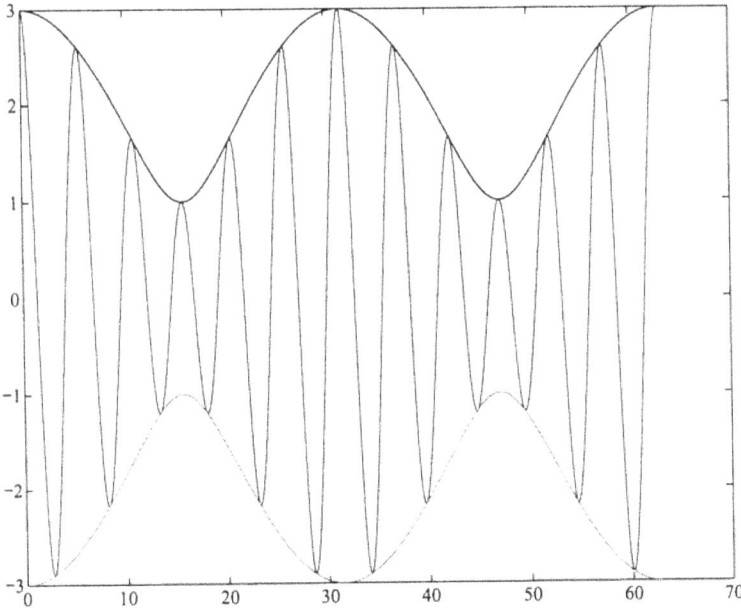

Figure 1.7   The signal $x(t) = \cos(t) + 2\cos(1.2t)$ and its amplitude obtained with the Hilbert transform.

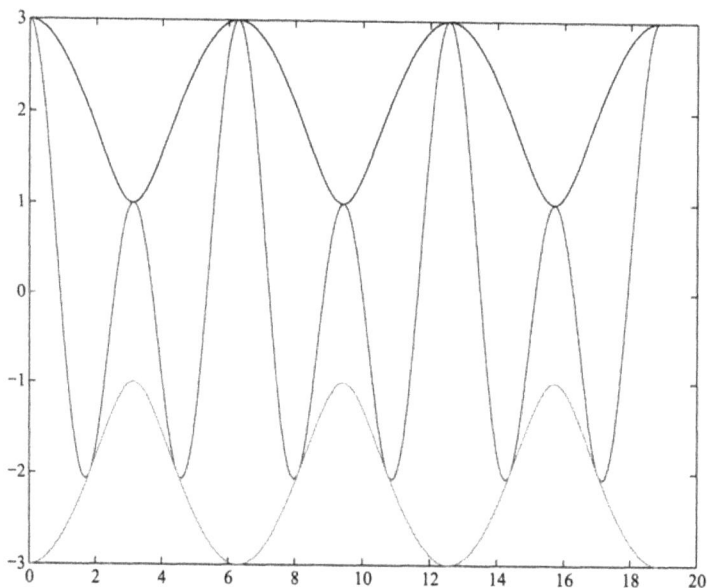

Figure 1.8   The signal $x(t) = \cos(t) + 2\cos(2t)$ and its amplitude obtained with the Hilbert transform.

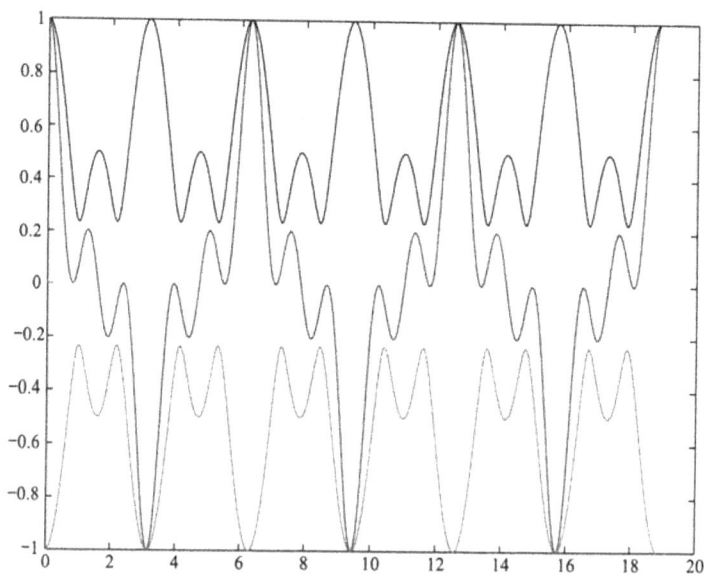

Figure 1.9   The signal $x(t) = \frac{1}{4}(2\cos(t) + \cos(3t) + \cos(5t))$ and its amplitude obtained with the Hilbert transform.

Example 1.4. *Let* $x(t) = \cos(t) + 2\cos(2t)$. *Its instantaneous amplitude* *is shown in Figure 1.8. It is observed that the upper envelope is physically meaningful but the lower envelope does not.*

Example 1.5. *Let $x(t) = \frac{1}{4}(2\cos(t) + \cos(3t) + \cos(5t))$. Its instantaneous amplitude is shown in Figure 1.9. It is easy to see that neither its upper envelope nor lower envelope provides physically meaningful demodulation.*

1.5   Paradoxes regarding the instantaneous frequency defined by the analytic signal

The instantaneous frequency defined by the Analytic Signal is a commonly accepted definition. However, it has some bewildering problems.

Example 1.6. *Suppose $s(t) = A_1 e^{i\omega_1 t} + A_2 e^{i\omega_2 t}$, where $\omega_1, \omega_2 > 0$, then take $s(t) = A(t)\cos\theta(t)$ where*

$$A(t) = (A_1^2 + A_2^2 + 2A_1 A_2 \cos((\omega_1 - \omega_2)t))^{1/2},$$

$$\theta(t) = \arctan\frac{A_1\sin(\omega_1 t) + A_2\sin(\omega_2 t)}{A_1\cos(\omega_1 t) + A_2\cos(\omega_2 t)}.$$

Therefore,

$$\theta'(t) = \frac{1}{2}(\omega_2 - \omega_1)\left[1 + \frac{A_2^2 - A_1^2}{A^2(t)}\right].$$

Figure 1.10(a) shows the waveforms of the signal $s(t) = 0.2\cos(10t) + \cos(20t)$. Its instantaneous frequency defined by the Analytic Signal is shown in Figure 1.10(c). Figure 1.10(b) shows the waveforms of the signal $s(t) = -1.2\cos(10t) + \cos(20t)$. And its instantaneous frequency defined by the Analytic Signal is shown in Figure 1.10(d).

From these two examples, we find out that [10]:

(1) Instantaneous frequency may not be one of the frequencies in the spectrum. A frequency existing at some time may not be in the final spectrum;

(2) Although the original signal consists of only two different frequencies, the instantaneous frequency may be continuous and range over infinite number of values;

(3) Although the spectrum of the analytic signal is zero for negative frequencies, the instantaneous frequency may be negative;

(4) For a bandlimited signal the instantaneous frequency may go outside the band;

(5) The instantaneous frequency defined by the analytic signal depends on the Hilbert transform of a signal, so it is not a local variable. This conflicts with our original intention of the instantaneous frequency.

Figure 1.11 shows how the negative instantaneous frequency occurs.

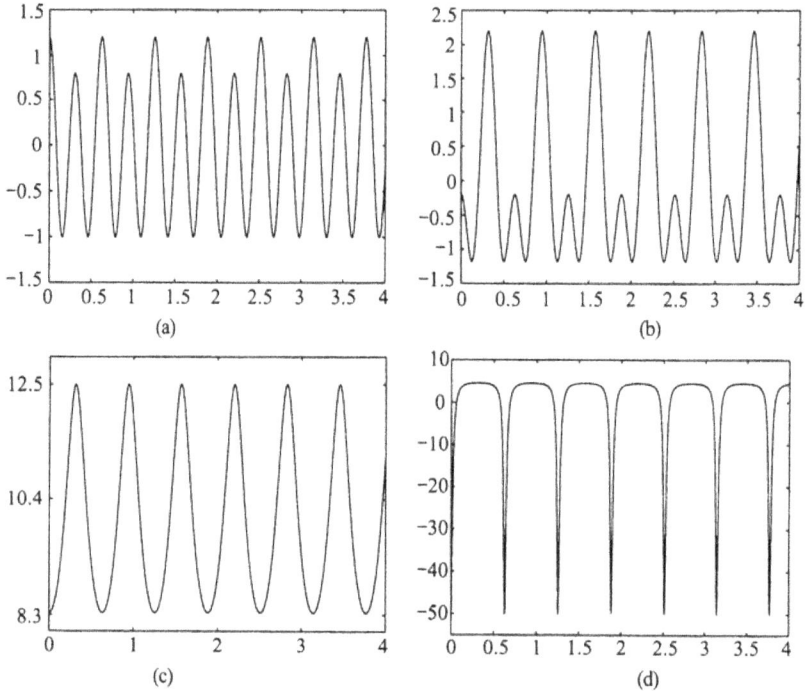

Figure 1.10   (a) The signal $s(t) = 0.2\cos(10t) + \cos(20t)$; (b) The signal $s(t) = -1.2\cos(10t) + \cos(20t)$; (c) Instantaneous frequency of the signal (a); (d) Instantaneous frequency of the signal (b).

It is seen that the instantaneous frequency defined by the analytic signal does not always give a physically meaningful characterization for an arbitrary signal. Physically, a signal may have various frequencies at a time, so the instantaneous frequency is just a synthesis of these frequencies. It is proved that (see[10])

$$\int \omega |\hat{s}(\omega)|^2 d\omega = \int \theta'(t)|s(t)|^2 dt.$$

Therefore, the instantaneous frequency does not provide a complete characterization of the frequencies of a signal. A signal may contain many different frequencies at any instant.

A signal which has only one frequency component at any time is called a 'monocomponent' signal.

It is just an intuitive definition because we really do not know what is the precise meaning that a signal has only one frequency component at any time.

Now we have two important questions: (1) What is the definition of

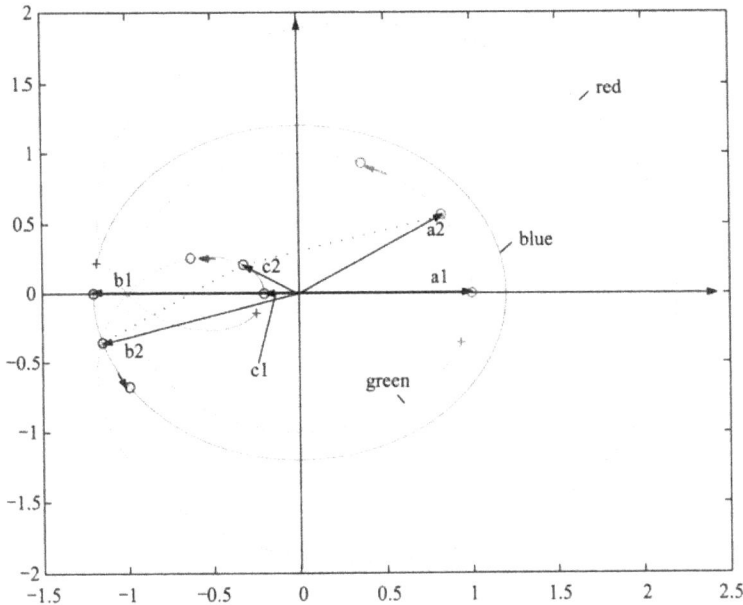

Figure 1.11   The tracks of analytic signals $z_1(t) = -1.2e^{i10t}$ (blue), $z_2(t) = e^{i20t}$ (green) and $z(t) = -1.2e^{i10t} + e^{i20t}$ (red), where t increases from 0 to 0.61. Denote $\mathbf{a1} := z_1(0)$, $\mathbf{b1} := z_2(0)$, $\mathbf{c1} := z(0) = \mathbf{a1} + \mathbf{b1}$ for $t = 0$. It is observed that when t increases from 0 to 0.03, $z_1(t)$ varies from $\mathbf{a1}$ to $\mathbf{a2} := z_1(0.03)$, $z_2(t)$ from $\mathbf{b1}$ to $\mathbf{b2} := z_2(0.03)$, and correspondingly $z(t)$ from $\mathbf{c1}$ to $\mathbf{c2} := z(0.03) = \mathbf{a2} + \mathbf{b2}$. In this process, the phase of signal $z(t)$ decreases (clockwise) and the negative instantaneous frequency occurs.

a monocomponent signal, and (2) how to decompose a given signal into a summation of monocomponent signals.

2   Hilbert-Huang transform

2.1   Intrinsic mode functions

In 1998, Huang *et al.* introduced the concept of Intrinsic Mode Function (IMF) that admits well-behaved Hilbert transform.

Definition 2.1. *A function $f(t)$ is defined to be an Intrinsic Mode Function (IMF) of a real variable t, if it satisfies two characteristic properties:*

(a) *In the whole data set, the number of extrema and the number of zero crossings must either equal or differ at most by one;*

(b) *At any point, the mean value of the envelope defined by the local maxima and the envelope defined by the local minima is zero.*

In Figure 2.1, (a) shows a harmonic signal $s(t) = \cos(2\pi t)$, (b) is the figure of a signal given by

$$
s(t) := \begin{cases} (t + \frac{1}{3})^2 \cos(2\pi(t+4)^2), & t \in \left[-3, -\frac{1}{3}\right), \\ 0, & t \in \left[-\frac{1}{3}, \frac{1}{3}\right], \\ (t - \frac{1}{3})^2 \cos(2\pi(t+3)^2), & t \in \left[\frac{1}{3}, 3\right], \end{cases} \qquad (2.1)
$$

and (c) shows a piecewise line signal. Obviously, all the signals are IMFs.

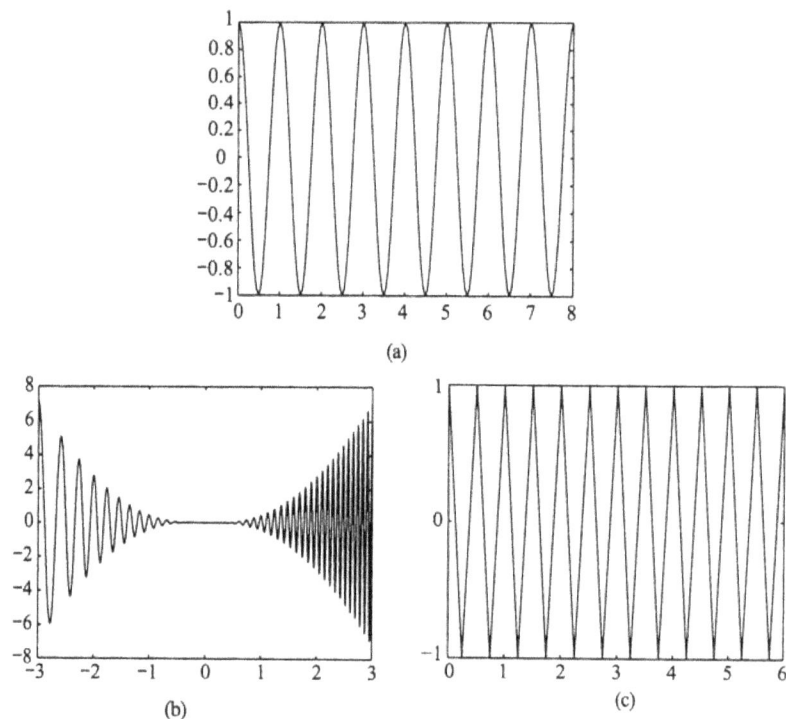

Figure 2.1   (a) A harmonic signal $s(t) = \cos(2\pi t)$; (b) The signal defined by Equation (2.1); (c) A piecewise line signal.

2.2   Empirical mode decomposition

Since IMFs admit well-behaved Hilbert transforms, Huang et al. consider the IMF as a kind of monocomponent signals. In [14], Huang et al. proposed a method called Empirical Mode Decomposition (EMD) to decompose any complicated data set into a collection of IMFs. Below is the EMD algorithm [14].

Algorithm 2.2. *Let $X(t)$ be the original signal.*

Step 1 *Initialization: let $r_0(t) = X(t)$, $i = 1$.*

Step 2 *The ith IMF can be extracted by the following iteration:*

    (a) *Set $h_0(t) = r_{i-1}(t)$, $j = 1$;*

    (b) *Identify all the extrema of $h_{j-1}(t)$;*

    (c) *Connect all the local maxima of $h_{j-1}(t)$ by a cubic spline curve as the upper envelope, designated as $u_{j-1}(t)$. Repeat the procedure for the local minima to produce the lower envelope $l_{j-1}(t)$;*

    (d) *Compute the mean $m_{j-1}(t) = \frac{u_{j-1}(t) + l_{j-1}(t)}{2}$;*

    (e) *Let $h_j(t) = h_{j-1}(t) - m_{j-1}(t)$;*

    (f) *If $h_j(t)$ is an IMF, then let $imf_i(t) = h_j(t)$; otherwise, set $j = j + 1$ and go back to (b).*

Step 3 *Let $r_i(t) = r_{i-1}(t) - imf_i(t)$.*

Step 4 *If there are more than two extrema of $r_i(t)$, then set $i = i + 1$ and go back to Step 2; otherwise, the algorithm ends and $r_i(t)$ is treated as a residue.*

With EMD, any complicated data set can be decomposed into a finite and often small number of Intrinsic Mode Functions (IMFs) whose instantaneous frequency defined by means of the Analytic Signal method should provide physically meaningful characterizations. After performing the Hilbert transform on each IMF component, we can express the data in the following form:

$$s(t) = \text{Re} \sum_{k=1}^{n} [imf_k(t) + iH imf_k(t)] = \text{Re} \sum_{k=1}^{n} a_k(t) e^{i\theta_k(t)}. \quad (2.2)$$

Equation (2.2) provides both the amplitude and the frequency of each component. If expanded with Fourier representation the signal can be expressed as

$$s(t) = \text{Re} \sum_{k=-\infty}^{\infty} a_k e^{ikt}, \quad (2.3)$$

where a_k is a constant for any integer k. The contrast between Equation (2.2) and Equation (2.3) is clear: EMD provides a more general expansion than Fourier.

Because of its excellence, EMD has been successfully used to analyze many natural signal. Let us cite a example from [14] here, which is a

tidal signal and its decomposition result by EMD are shown in Figure
2.2. The wave data were collected from the tidal gauge, located inside Kahului Harbour, Maui, for five days from October 4-9, 1994. On
October 5, tsunami-induced waves arrived at the site and created wa-
ter level changes of a magnitude comparable to that of the tidal signal.
Although the tidal data are traditionally analyzed with Fourier expan-

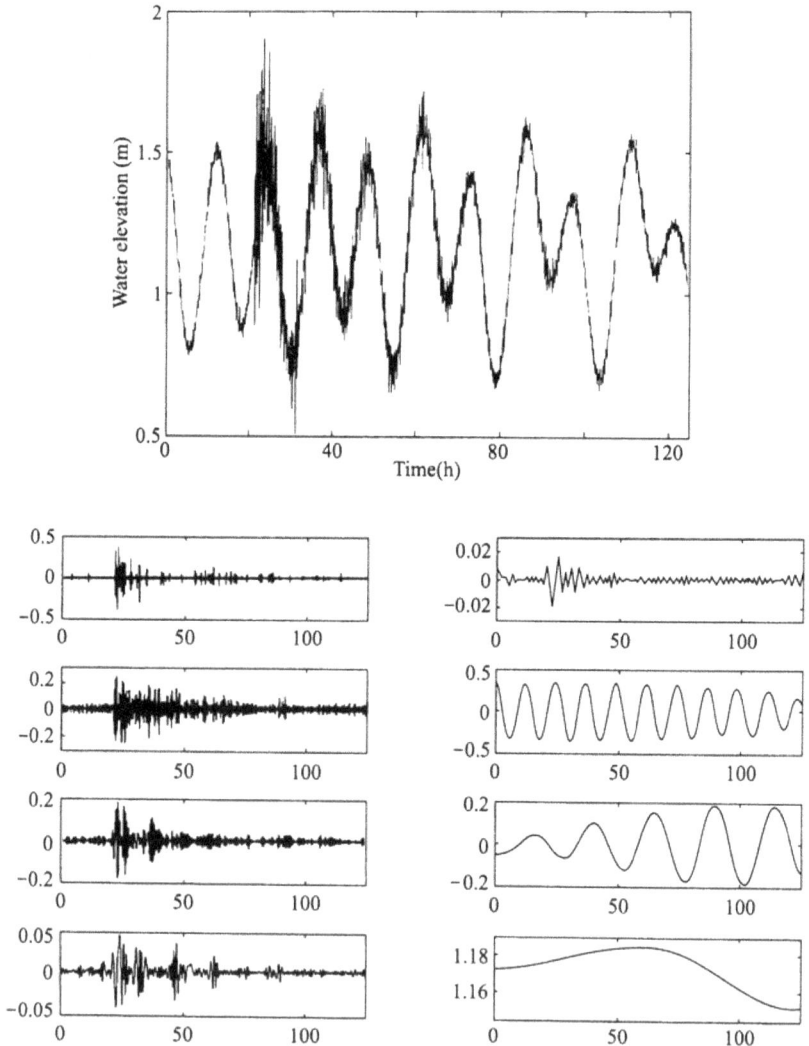

Figure 2.2   Top: The tidal data collected inside the Kahului Harbour, Maui, from
October 4-9, 1994; Bottom: The seven IMF components and a residue of the tidal
data obtained by the EMD method.

sion the added tsunami waves are transient. The combination, therefore, makes the whole time series non-stationary. Filtering can not remove the tsunami signal cleanly, for the transient data and the tide will have many harmonic components in the same frequency range. Because the time scales of the tide and the tsunami waves are so different, the eight IMF components can be easily divided into two groups: the high-frequency signal representing the tsunami-induced waves; and the last three low-frequency components representing the tide. After the IMFs were used to reconstitute the two separate wave motions, the raw data, the tidal component and the tsunami-induced waves are plotted together in Figure 2.3(a). Here, the EMD serves as a filter to separate the tide and the transient tsunami without any ambiguity (cf. [14]).

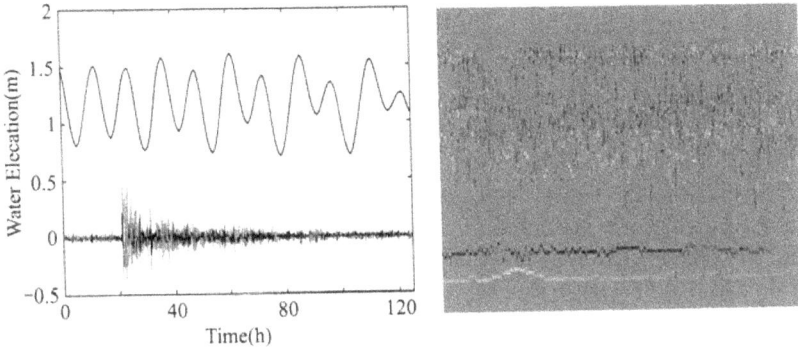

Figure 2.3   (a) Separation of the tides and the tsunami waves by the IMF components as shown in Figure 2.2: the raw data are shown in the dotted line; the tides represented by the sum of the last three IMFs shown in the solid line; and the tsunami waves represented by the sum of the first five IMFs shown in the thick solid line at the bottom. (b) The Hilbert spectrum of the tidal signal shown in Figure 2.2.

2.3   Hilbert spectrum

Equation (2.2) also enables us to represent the amplitude and the instantaneous frequency as functions of time in a three-dimensional plot, in which the amplitude can be contoured on the frequency-time plane. This frequency-time distribution of the amplitude is designated as the Hilbert spectrum $H(\omega, t)$ whose precise expression is given as follows.

$$H(\omega, t) = \begin{cases} 0, & \text{if } J_{\omega,t} = \emptyset, \\ \sum_{k \in J_{\omega,t}} a_k(t), & \text{if } J_{\omega,t} \neq \emptyset, \end{cases} \qquad (2.4)$$

where

$$J_{\omega,t} = \{k \mid 0 \leq k \leq n, \ \theta_k'(t) = \omega\}.$$

The Hilbert spectrum can give us a full energy-frequency-time distribution of the data. It would be ideal for nonlinear and non-stationary data analysis. To show its effectiveness, Figure 2.3(b) gives the Hilbert spectrum of the tidal signal shown in Figure 2.2, from which the arrival time and the frequency change of the tsunami waves are clearly shown in this energy-frequency-time distribution. Besides the clear dispersion properties of the tsunami waves, there are two more interesting new observations: first, the variations of the tsunami wave frequency are phase locked with the tidal cycle: second, the tsunami waves in the harbor lasted many tidal cycles, with a frequency of half a cycle per hour [14].

The combination of the EMD and the Hilbert spectrum is called the Hilbert-Huang Transform (HHT). Since the wavelet analysis is also a popular time-frequency distribution which can achieve excellent localization features both in time and frequency, we compare the wavelet analysis with HHT here.

The wavelet analysis is an effective time-frequency analysis method for non-stationary signals. The basic wavelet function, $\psi_{j,k}(t)$, which is the dilation and translation of a mother wavelet function, possesses an intrinsic time-frequency structure. This makes the wavelet analysis locate a signal both in time and frequency domains adaptively. However, the wavelet approach is essentially an adjustable windowed Fourier spectral analysis. Once the mother wavelet function is chosen, it will be used to analyze all the data. So the time-frequency structure of the mother wavelet function can not be changed according to the oscillations of the signal adaptively at different time.

EMD can decompose any complicated signal into a finite number of IMFs. Each IMF is a physically meaningful time-frequency structure of the original signal. Furthermore, the IMFs, i.e. the basis of the decomposition, are derived from the original signal adaptively. Therefore, the EMD is truly an adaptive and posteriori method [14].

3   Some relevant questions and our recent researches

The EMD is an adaptive data decomposition. Any component obtained from EMD can be regarded as a basic atom for time-frequency analysis. The efficiency of HHT has been tested by many applications.

However, it is still an empirical algorithm and the mathematical foundation has not been established. Many relevant theoretic questions are still open. This section will summarizes some of our recent works in this field.

3.1   Empirical AM/FM demodulation and an improvement

3.1.1   Empirical AM/FM demodulation

IMFs obtained by EMD do not always provide physically meaningful demodulation through Hilbert transform. An example in Figure 3.1 illuminates this phenomena (c.f.[36]).

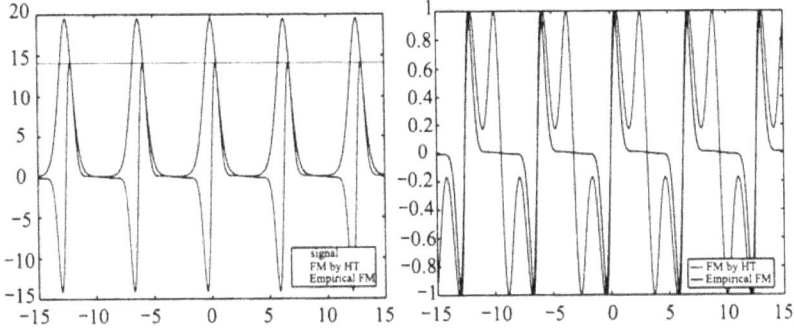

Figure 3.1   AM and FM from empirical method and HT. The signal $s(t) = e^{\epsilon \cos t} \sin(\epsilon \sin t))$ ($\epsilon = 2.97$) is from [36].

Therefore, Huang *et al.* proposed the empirical AM/FM demodulation [15]. It is an iterative normalization scheme enabling any IMF to be separated empirically and uniquely into envelope (AM) and carrier (FM) parts, which eschews the Hilbert transform totally. This algorithm is simply presented as follows: Firstly, for a given IMF data set $s(t)$, identify all the local maxima of $|s(t)|$. Then all these maxima points are connected with a cubic spline curve. This spline curve is designated as the empirical envelope of the data, $e_1(t)$. Do the first normalization procedure to $s(t)$:

$$y_1(t) = \frac{s(t)}{e_1(t)},$$

and $y_1(t)$ is called the first normalized data. If $|y_1(t)| \leq 1$, the normalization is complete: otherwise, $|y_1(t)|$ still has some amplitudes higher than unity occasionally. Then, the normalization procedure must be implemented repeatedly, with $e_2(t)$ defined as the empirical envelope of $y_1(t)$ and so on as,

$$y_2(t) = \frac{y_1(t)}{e_2(t)}.$$

After the nth iteration, when all of the values of $y_n(t)$ are less than or equal to unity, the normalization is complete. It is designated as the empirical FM (Frequency Modulation) part of the data, $F(t)$, as

$$F(t) := y_n(t) = \cos\phi(t).$$

With the FM part determined, the AM (Amplitude Modulation) part, $A(t)$, can be simply obtained as,

$$A(t) := e_1(t)\cdots e_n(t).$$

Therefore $s(t)$ can be written as

$$s(t) = A(t)F(t) = A(t)\cos\phi(t).$$

Having obtained the empirical FM part, $\cos\phi(t)$, the empirical IF can be computed simply by the derivative of the phase function,

$$\omega(t) := \frac{\phi'(t)}{2\pi}.$$

The above demodulation proposed by Huang *et al.* in [15] does not depend on the Hilbert transform, and is designated as the empirical AM/FM demodulation. Experiments shows its validity in [15].

3.1.2   Riding waves and an improvement to the algorithm

It is found in [48] that with the above AM/FM demodulation, the empirical FM part may contain riding waves. A typical example is given as follows. Let

$$s(t) = \exp(20\cos(t))\sin(20\sin(t)) \qquad (|t| \le 12.55) \qquad (3.1)$$

and set the sampling rate as 1000Hz in our numerical experiment. Applying EMD to $s(t)$ we get three IMFs and one residue, as shown in Figure 3.2.

Let us demodulate IMF_1, the first IMF of $s(t)$, with the empirical AM/FM demodulation. $|IMF_1|$ and its empirical envelope are shown in Figure 3.3. It is easy to see that undershoots appear in the neighborhood of $t = -11, -9, -3, 3, 9$ and 11. The bottom of Figure 3.3 is a magnified display of the undershoot near $t = -3$.

After five rounds of iterative normalization, we get the empirical FM part of IMF_1, $\cos\phi_e(t)$, as shown in Figure 3.4, in which two riding waves occur in the neighborhood of $t = -7$ and $t = 7$ respectively.

It is also found that the empirical FM part may still contain riding waves even though no undershoot occurs in the above iterative normalization [48].

In order to get the physically meaningful instantaneous frequency and amplitude, a new technique called Riding Wave Turnover Method

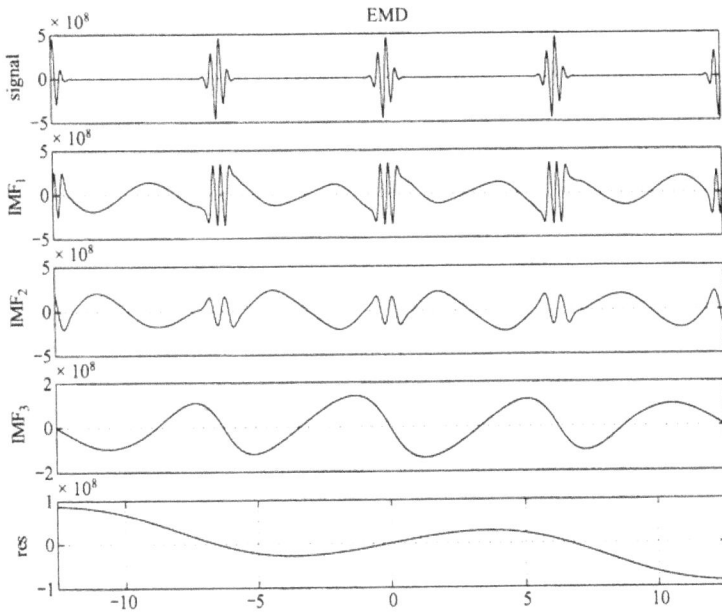

Figure 3.2   Top: The original signal $s(t)$; Middle: Its three IMFs; Bottom: The residue.

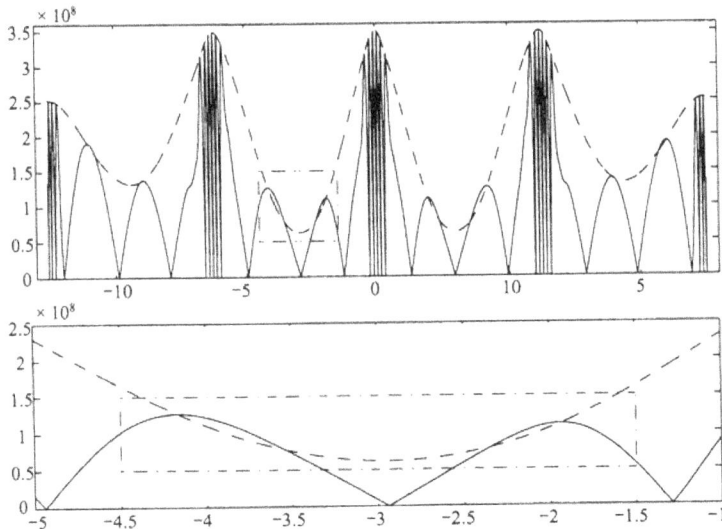

Figure 3.3   Top: $|IMF_1|$ (solid line) and its empirical envelope (dashed line). There are two undershoots in the dashdotted rectangle near $t = -3$; Bottom: A magnified display of the dashdotted rectangle.

(RWTM) is developed by Yang, Yang, Qing and Huang to eliminate the riding waves in the demodulation [48]: Assume that there is a riding

158 Lihua Yang

wave in the signal $s(t)$ whose endpoints are $(t_a, s(t_a))$ and $(t_b, s(t_b))$ respectively. Let $l(t)$ be the linear segment connecting these two endpoints. Then replacing $s(t)$ by $2l(t) - s(t)$ for all $t \in [t_a, t_b]$, as shown in Figure 3.5 where $s(t)$ is displayed with solid line and $2l(t) - s(t)$ with dashed line, we eliminate the riding wave. Applying RWTM to all riding waves will eliminate all the riding waves.

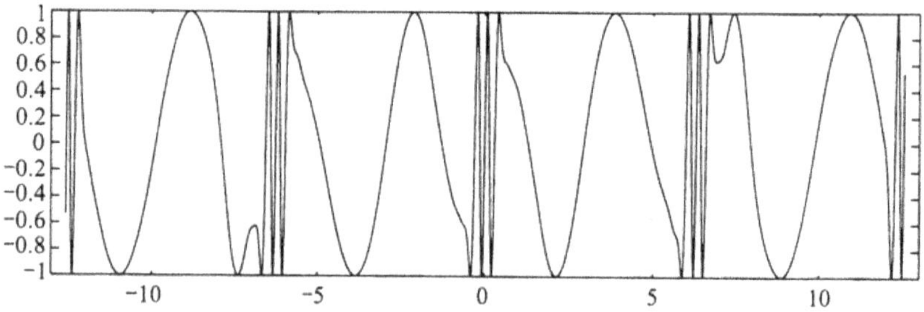

Figure 3.4   The empirical FM part $\cos \phi_e(t)$ of IMF$_1$ after five rounds of iterations. Riding waves appear in the neighborhood of $t = -7$ and $t = 7$ respectively.

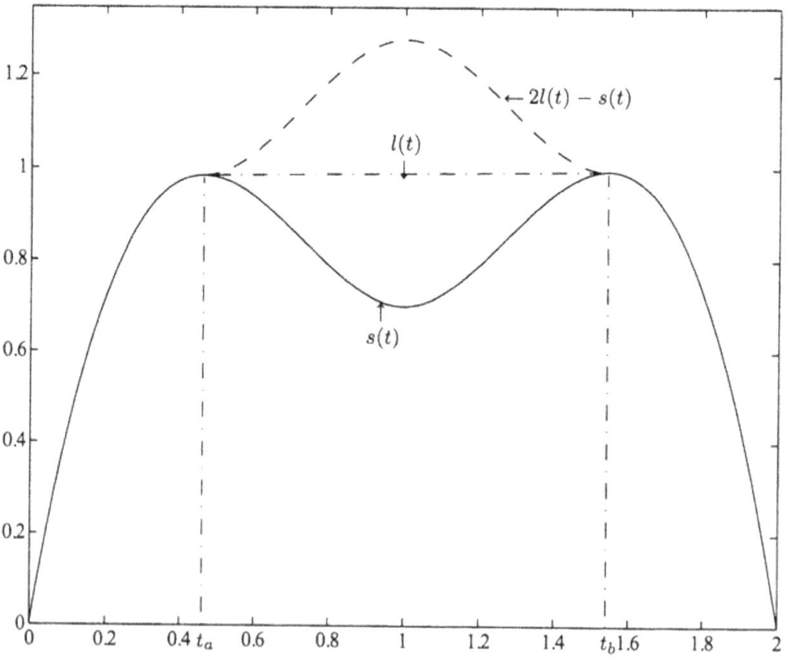

Figure 3.5   The original signal containing a riding wave (solid line) and the result of turning the riding wave upwards (dashed line).

Combining the RWTM with the empirical AM/FM demodulation, an improved demodulation method is presented in [48], which is called *riding wave turnover-empirical AM/FM demodulation.* The algorithm is formulated as follows:

Algorithm 3.1 (Riding wave turnover-empirical AM/FM demodulation).

Let $s(t)$ be the original signal.

Step 1: *Apply the empirical AM/FM demodulation to $s(t)$ to obtain its empirical FM part $F(t)$.*

Step 2: *If $F(t)$ contains riding waves then eliminate all the riding waves by applying RWTM to $F(t)$; denote the resultant signal by $s(t)$ and go to Step 1; otherwise, go to Step 3.*

Step 3: *$F(t)$, which contains no riding waves, is the final empirical FM part of $s(t)$. The program ends.*

Applying the riding wave turnover-empirical AM/FM demodulation to the IMF₁ of $s(t)$ discussed above we get the result shown at the bottom of Figure 3.6. To make the comparison more clear, IMF₁ and the

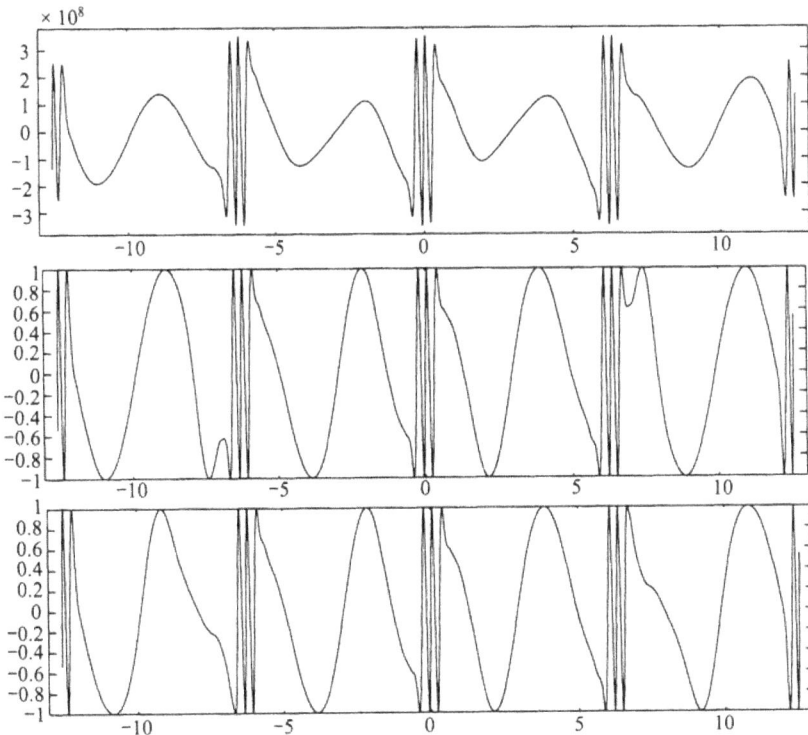

Figure 3.6   Top: Imf₁ of $s_1(t)$; Middle: Its empirical FM part extracted by the empirical AM/FM demodulation; Bottom: Its empirical FM part extracted by our improved method.

FM part extracted by the empirical AM/FM demodulation (c.f. Figure 3.4) are shown again at the top and in the middle of Figure 3.6. It is seen that riding waves are eliminated in the FM part extracted by our improved method. Physically, it is observed easily that a serious deformation happens in the local interval in the empirical FM part extracted by the empirical AM/FM demodulation owing to the riding wave (see the middle of Figure 3.6). However, that extracted by our improved method possesses almost the same waveform as the original signal (see the bottom of Figure 3.6).

3.2   Hilbert transform and the Bedrosian identity

Intrinsic mode functions are assumed to admit better-behaved Hilbert transforms. Let $s(t)$ be an IMF, with Hilbert transform, its instantaneous amplitude $\rho(t)$ and phase $\theta(t)$ are solved as

$$s(t) + iHs(t) = \rho(t)e^{i\theta(t)},$$

which means that $\rho(t)$ and $\theta(t)$ should satisfy

$$H[\rho(t)\cos\theta(t)] = \rho(t)\sin\theta(t). \tag{3.2}$$

However, it is observed that under proper conditions the following equality

$$H[\rho(t)\cos\theta(t)] = \rho(t)H\cos\theta(t) \tag{3.3}$$

holds, which means that (3.2) is true if and only if $H\cos\theta(t) = \sin\theta(t)$. Therefore, demodulation equation (3.2) can be reduced to a question of frequency demodulation of a unitary amplitude signal [15, 50]. Equation (3.3) is a very important equality for demodulation of signals in the non-stationary signal processing, whose general form is

$$H(fg) = fHg. \tag{3.4}$$

Equation (3.4) was first studied by Bedrosian in 1963 and called the Bedrosian identity in honor of him [4]. Later, Nuttall and Bedrosian (1966), and Brown (1974) obtained more general sufficient conditions [7, 25]. In 1986, Brown established the first necessary and sufficient condition in the time domain and a parellel result in the frequency domain for the Bedrosian identity [8] to be valid, which we state as follows:

(a) If $f, g \in L^2(\mathbb{R})$ are bounded on $\mathbb{R}$, then $H(fg) = fH(g)$ if and only if

$$H(f_+(t)g_-(t)) = if_+(t)g_-(t), \quad H(f_-(t)g_+(t)) = -if_-(t)g_+(t), \tag{3.5}$$

or

$$\hat{f}_+ * \hat{g}_-(\omega) = 0 \text{ a.e. } \omega \in \mathbb{R}_+, \quad \hat{f}_- * \hat{g}_+(\omega) = 0 \text{ a.e. } \omega \in \mathbb{R}_-, \tag{3.6}$$

where '$*$' denotes the convolution operator, $g_+(t) = (\hat{g}(\omega)\chi_{\mathbb{R}_+})\check{}\,(t)$ and $g_-(t) = (\hat{g}(\omega)\chi_{\mathbb{R}_-})\check{}\,(t)$ with $\mathbb{R}_+ = (0, \infty)$ and $\mathbb{R}_- = (-\infty, 0)$. Hereafter, $\check{f}$ denotes the inverse Fourier transform of f and χ_E the characteristic function of set E.

Recently, as the advent of the Hilbert-Huang transform, the Bedrosian identity received much attention again. It was proved in [49, Theorem 2.4] and [42, Theorem 2.3] that (3.6) and (3.5) are still necessary and sufficient conditions for the Bedrosian identity even if the functions f and g in (b) are not bounded. Another new necessary and sufficient condition was proposed by Xu and Yan in [43]:

(b) If $f, f', g \in L^2(\mathbb{R})$, then $H(fg) = fH(g)$ if and only if

$$\int_0^1 dt \int_{\mathbb{R}} \frac{\omega}{t^2} e^{-2i\pi x \omega(t-1)/t} \hat{f}\left(\frac{\omega}{t}\right) \hat{g}(-\omega) d\omega = 0. \qquad (3.7)$$

The result was improved by Tan, Yang and Huang in [42] as follows:

Theorem 3.2. *Let $f, g \in L^2(\mathbb{R})$. If $\int_A |\hat{g}(\lambda)\hat{f}(\omega - \lambda)| d\omega d\lambda < \infty$, where $A := [\mathbb{R}_- \times \mathbb{R}_+] \cup [\mathbb{R}_+ \times \mathbb{R}_-]$, then $H(fg) = fHg$ if and only if (3.7) holds.*

Corollary 3.3. *Let $f, g \in L^2(\mathbb{R})$ satisfying $|x|^\alpha \hat{f}(x), |x|^{1-\alpha} \hat{g}(x) \in L^2(\mathbb{R})$ for some $\alpha > \frac{1}{2}$. Then $H(fg) = fHg$ if and only if (3.7) holds.*

It is easy to see that Xu and Yan's result is the special case corresponding to $\alpha = 1$.

The above results on the Bedrosian identity always assume that both f and g are in $L^2(\mathbb{R})$. However, if both of them are periodic or only g is periodic, does the Bedrosian identity still holds? In these cases, the Hilbert transform should be replaced by the following circular Hilbert transform for periodic functions defined by

$$\tilde{H}f(x) := \frac{1}{\pi}\text{p.v.} \int_{-T/2}^{T/2} \frac{f(t-\tau)}{2\tan\frac{\tau}{2}} d\tau, \qquad f \in L_T^p, \qquad (3.8)$$

where, p.v. is the Cauchy principal and L_T^p ($1 \le p < \infty$) is the space of all the T-periodic function f satisfying

$$\|f\|_{L_T^p} := \left(\int_0^T |f(x)|^p dx\right)^{1/p} < \infty.$$

In the case that f is in $L^2(\mathbb{R})$ and g is periodic, the following Bedrosian identity was established by Tan, Yang and Huang [42]:

Theorem 3.4. *Let $f \in \mathcal{L}_T^2(\mathbb{R})$, $g \in L_T^2$. Then $H(fg) = fHg$ if and only if*

$$
\begin{cases}
c_0(g)\hat{f}(\omega) + 2 \displaystyle\sum_{k=-\infty}^{-1} c_k(g)\hat{f}(\omega - \frac{2\pi}{T}k) = 0, & \text{a.e. } \omega \in (0, \infty), \\
c_0(g)\hat{f}(\omega) + 2 \displaystyle\sum_{k=1}^{\infty} c_k(g)\hat{f}(\omega - \frac{2\pi}{T}k) = 0, & \text{a.e. } \omega \in (-\infty, 0)
\end{cases}
$$

holds, where $\mathcal{L}_T^2(\mathbb{R})$ for $T > 0$ is defined as the space of all the Lebesgue integrable functions on $\mathbb{R}$ satisfying

$$
|f|_T := \left\| \sum_{k \in \mathbb{Z}} |f(\cdot + kT)| \right\|_{L_T^2} < \infty.
$$

The case that both f and g are in $L_{2\pi}^2$ was studied by Qian, Yu and Zhang [32, 49]. However, if the case that f and g are periodic but with different periods is more often encountered in applications, such as the beat waves in signal processing [23] (Figure 3.7). Below is a result set up by Tan, Yang and Huang for this case [42].

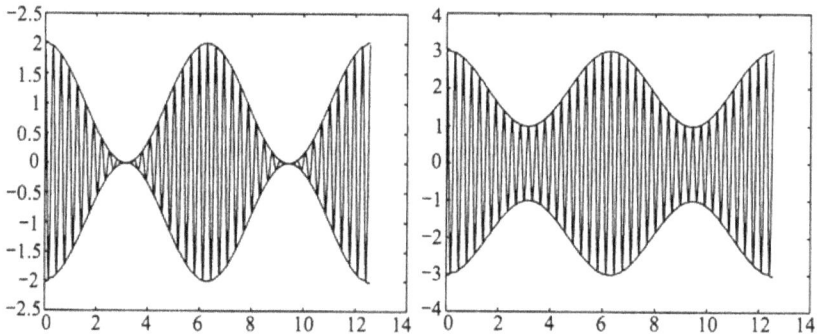

Figure 3.7   Beat waves. Left: $s(t) = (1 + \cos(20t)) \cos t$; Right: $s(t) = (2 + \cos(20t)) \cos t$.

Theorem 3.5. *Let $T_1, T_2 > 0$ have a common multiple $T = nT_1 = mT_2$ and $f \in L_{T_1}^2, g \in L_{T_2}^2$. Then $H(fg) = fHg$ if and only if*

$$
\sum_{k \in \mathbb{Z}} [\operatorname{sgn}(l) - \operatorname{sgn}(k)] c_k^{T_2}(g) c_{\frac{l-km}{n}}^{T_1}(f) = 0 \quad \text{for all } l \in \mathbb{Z}.
$$

Finally, the Bedrosian identity for $H(fg) = fHg$ for $f \in L^p(\mathbb{R})$ and $g \in L^q(\mathbb{R})$ with $1 \leq p, q \leq \infty$ was studied by Yang and Zhang [44]. Denote

$$
\mathbf{H}^p(\mathbb{R}) := \{f | f, Hf \in L^p(\mathbb{R})\} \ (1 \leq p < \infty) \quad \text{and} \quad \mathbf{H}^\infty(\mathbb{R}) := L^\infty(\mathbb{R}).
$$

Using the facts that $\text{BMO}(\mathbb{R}) = (\mathbf{H}^1(\mathbb{R}))'$, where $\text{BMO}(\mathbb{R})$ is the space of all the bounded mean oscillation functions (see [40]), and $H^2 f = -f$ $(\forall f \in \mathbf{H}^1(\mathbb{R}))$ the Hilbert transform of $f \in \text{BMO}(\mathbb{R})$ can be defined by the following equation $\langle Hf, g \rangle = -\langle f, Hg \rangle$ $(\forall g \in \mathbf{H}^1(\mathbb{R}))$, which gives that $Hf \in \text{BMO}(\mathbb{R})$ if $f \in \text{BMO}(\mathbb{R})$. Since $L^\infty(\mathbb{R})$, the space of all the essentially bounded functions on $\mathbb{R}$, is a subspace of BMO, the Hilbert transform makes sense for $f \in L^\infty(\mathbb{R})$. The Bedrosian identity was generalized by Yang and Zhang as follows in [44].

Theorem 3.6. *Let $f \in \mathbf{H}^p(\mathbb{R}), g \in \mathbf{H}^q(\mathbb{R})$ with $p, q \geq 1$, $1/p + 1/q \leq 1$. Then f, g satisfy the Bedrosian identity $H(fg) = fHg$ if and only if*

$$\text{supp}(f_+ g_-)\hat{} \subseteq \mathbb{R}_- \quad and \quad \text{supp}(f_- g_+)\hat{} \subseteq \mathbb{R}_+, \tag{3.9}$$

where $h_+ := \frac{1}{2}(h + iHh)$ and $h_- := \frac{1}{2}(h - iHh)$.

Corollary 3.7. *If $f \in \mathbf{H}^p(\mathbb{R}), g \in \mathbf{H}^q(\mathbb{R})$ with $p, q \geq 1$, $1/p + 1/q \leq 1$ satisfy either $(\text{supp}\hat{f}) \cup (\text{supp}\hat{g}) \subseteq \mathbb{R}_+$ or $(\text{supp}\hat{f}) \cup (\text{supp}\hat{g}) \subseteq \mathbb{R}_-$ then the Bedrosian identity holds.*

Corollary 3.8. *Let $f \in \mathbf{H}^p(\mathbb{R})$ and $g \in \mathbf{H}^q(\mathbb{R})$ with $p, q \geq 1$, $1/p + 1/q \leq 1$. If there exist nonnegative constants a, b such that $\text{supp}\hat{f} \subseteq [-a, b]$ and $\text{supp}\hat{g} \subseteq (-\infty, -b] \cup [a, \infty)$ then the Bedrosian identity holds.*

3.3   Hilbert transform on distribution spaces

Both H and $\tilde{H}$ defined by (1.4) and (3.8) respectively are called Hilbert transforms. A natural and interesting question is: what is the relation between them? If $s(t) = f(t) + g$ with $f \in L^2(\mathbb{R})$ and $g \in L^2_{2\pi}$, what is the Hilbert transform of s and how to find its instantaneous frequency and amplitude?

No literatures is reported on the research of the above questions. Up till now, many achievements have been made to extend the classical Hilbert transform to some generalized function spaces [5, 6, 26, 27, 28, 29]. Most of them (cf. [5, 6, 26]) on this topic is to extend the Hilbert transform to a preexistent distribution space by using the analytic representation of distributions. In [27], Hilbert transform is extended to $\mathscr{D}'$, the space of Schwartz distributions, directly with conjugate operator by introducing the topology on $H(\mathscr{D})$. With this extension, for any $f \in \mathscr{D}'$, its Hilbert transform Hf is in $H'(\mathscr{D})$, which is called a space of ultradistributions [27]. It is easily verified that $H'(\mathscr{D})$ is not a subspace of $\mathscr{D}'$ since $H\phi \notin \mathscr{D}$ for $\phi \in \mathscr{D}$ unless $\phi = 0$, which means that H is not closed in $\mathscr{D}'$.

Yang constructed a new distribution space $\mathscr{D}'_H$ and extended the Hilbert transform to it such that $\mathscr{D}'_H$ is a subspace of $\mathscr{D}'$ and H is a homeomorphism on $\mathscr{D}'_H$ [45]. $\mathscr{D}'_H$ is the dual of the direct sum space $\mathscr{D}_H := \mathscr{D} \dot{+} H(\mathscr{D})$, whose topology is defined as:

$$\forall \{\phi_n + H\psi_n\} \subset \mathscr{D}_H, \text{ define } \phi_n + H\psi_n \to 0 \ (in \ \mathscr{D}_H) \text{ if } \phi_n, \psi_n \to 0 \ (in \ \mathscr{D}). \tag{3.10}$$

Endowed with this topology, $\mathscr{D}_H$ becomes a topological vector space satisfying $H(\mathscr{D}_H) = \mathscr{D}_H$ and $H : \mathscr{D}_H \to \mathscr{D}_H$ is a continuous linear operator. Accordingly, H is a homeomorphism on $\mathscr{D}_H$ since $H^{-1} = -H$. It was shown in [45] that $L^p(\mathbb{R}) \subsetneq \mathscr{D}'_H$ and $\dot{L}^1_T \subsetneq \mathscr{D}'_H$, where $\dot{L}^1_T := \{f \in L^1_T | \int_0^T f(x)dx = 0\}$ and '$\mathscr{X} \subsetneq \mathscr{Y}$' means that the topological vector space $\mathscr{X}$ is continuously embedded into the topological vector space $\mathscr{Y}$ (c.f. [33]). Besides, it is also deduced that the Dirac impulse δ is in $\mathscr{D}'_H$. Using the conjugate operator, the classical Hilbert transform H is extended to the dual $\mathscr{D}'_H$ as follows:

Definition 3.9. *Let $H^* : \mathscr{D}'_H \to \mathscr{D}'_H$ be the conjugate operator of the classical Hilbert transform $H : \mathscr{D}_H \to \mathscr{D}_H$. Then $-H^* : \mathscr{D}'_H \to \mathscr{D}'_H$ is defined as the extension of H to the distribution space $\mathscr{D}'_H$, and denoted as H still if no confusion occurs.*

Theorem 3.10. *Let $f \in L^p(\mathbb{R})$ $(1 < p < \infty)$ (or: $f \in \dot{L}^1_T$). Then, Hf (or: $\tilde{H}f$), as the classical (circular) Hilbert transform, coincides with the extended one defined by Definition 3.9.*

4   Applications of Hilbert-Huang transform to pattern recognition

In this section we will give some applications of Hilbert-Huang transform to pattern recognition.

4.1   HHT-based detection of spindles in sleep EEGs

Sleep is a complicated physiological process. Generally, sleep consists of two phases: no-rapid eye movement (NREM) and rapid eye movement (REM). The NREM phase can be decomposed into 4 stages according to sleep depths [19]. A crucial clue for the sleep depth to be at the second or the third stages is that the sleep-spindles, whose frequencies are between 12 and 20Hz, take place in the EEG (electroencephalo-graph) [22].

Traditionally, sleep-spindles are detected visually by neurologists or sleep experts. Research on automated sleep analysis can be traced back to as early as the 1970s [38, 39, 31, 35, 12, 30, 34, 18]. In recent years, two novel algorithms for automated detection of spindles in sleep EEG were developed by using classical time-frequency analysis [22, 13]. Since EEGs are typically nonlinear and non-stationary signals and the duration of a sleep-spindle is usually very short, it is usually difficult to obtain satisfactory results in automated detection of sleep-spindles by using traditional time-frequency analysis. In [46] an new algorithm to detect spindles from sleep EEGs automatically is developed by Yang, Yang and Qi and the detection rate is higher than those developed in [22, 13].

In practice, a sleep EEG contains a large amount of data and is terribly time-consuming to calculate the Hilbert spectrum of a global sleep EEG signal. Thus, to save CPU time, a global EEG signal is divided into many short segments of of 1200 data (6 seconds for frequency of sampling = 200Hz). A 6 second segment, denoted by $X(t)$, is selected from a sleep EEG which is sampled when the sleep is at the 2nd stage of a NREM phase. It contains two sleep-spindles, marked by 'A' and 'B' respectively, as shown in Figure 4.1.

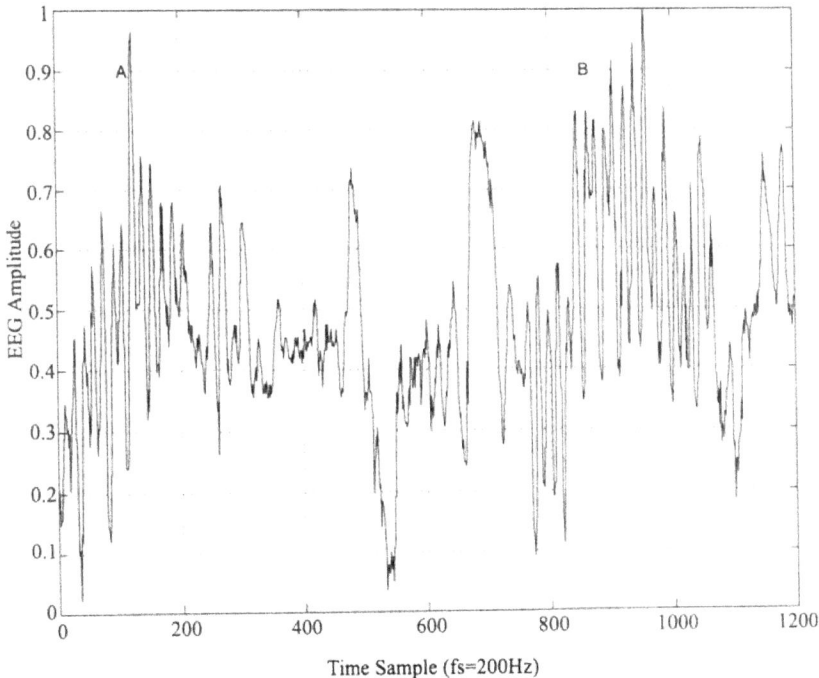

Figure 4.1   A sleep EEG segment of 6 seconds at the second sleep stage, in which two sleep-spindles are included as marked by 'A' and 'B' respectively.

For each segment, the sleep-spindles are detected by the following algorithm [46].

Algorithm 4.1. *Let $x(t)$ be a segment of data of length 1200. The sleep-spindles are detected as follows:*

Step 1 *Decompose $x(t)$ with EMD into IMFs, then remove the first*

IMF. For each other IMF, calculate its instantaneous frequency and instantaneous amplitude. Quantify the instantaneous frequency into integers between 1 and 100 Hz.

Step 2 *Compute the Hilbert spectrum $H(\omega, t)$; here it is a matrix of 100 rows and 1200 columns. Then normalize the amplitude of $H(\omega, t)$ linearly such that the values of $H(\omega, t)$ range from 0 to 255.*

Step 3 *Extract the 8th to 20th rows of $H(\omega, t)$ to form a sub-matrix, denoted by M, of 13 rows and 1200 columns.*

Step 4 *Calculate the maximum of each column of M to generate an array, $C = (C[1], \cdots, C[1200])$. It is an energy measure of the data on frequencies ranging from $8 \sim 20$ Hz at each local time. Then, define a smoothed version of C as:*

$$C_1[k] = \frac{1}{L} \sum_{i=k-L/2}^{k+L/2} C[i],$$

where L, an even integer, is the width of the smoothing window ($L = 50$ in the experiments of this paper) and the boundary extension is conducted as: $C[i] = C[1]$ for $i \leq 0$ and $C[i] = C[1200]$ for $i > 1200$.

Step 5 *Let T be a threshold. Then, we search $1 \leq k \leq 1100$ and $I \geq 100$ such that*

$$C_1[k + i - 1] \geq T \quad \text{for} \quad i = 1, 2, \cdots, I,$$

and

$$C_1[k + I] < T \quad \text{or} \quad k + I = 1200.$$

Then a sleep-spindle that starts at k and has duration I is detected. We set $T = 50$ in the experiments of this paper.

For the segment of a sleep EEG shown in Figure 4.1, by Algorithm 4.1 two spindles of a sleep EEG are detected as shown in Figure 4.2, in which the starting points, the durations, and the end points are marked by the dotted lines. The first starts at about the 20th datum (namely: the 0.1th second) with a duration of about 0.8s and the second starts at about the 750th datum (namely: the 3.75th second) with a duration of about 1.5s.

To test the detection algorithm, 100 segments, each of which consists of 1200 data (about 6 seconds with frequency of sampling 200Hz) and all of which contain 183 spindles, are selected from a sleep EEG database. The locations and durations of these sleep-spindles have been determined visually by experts. Let $X(t)$ be a sleep EEG segment which contains a

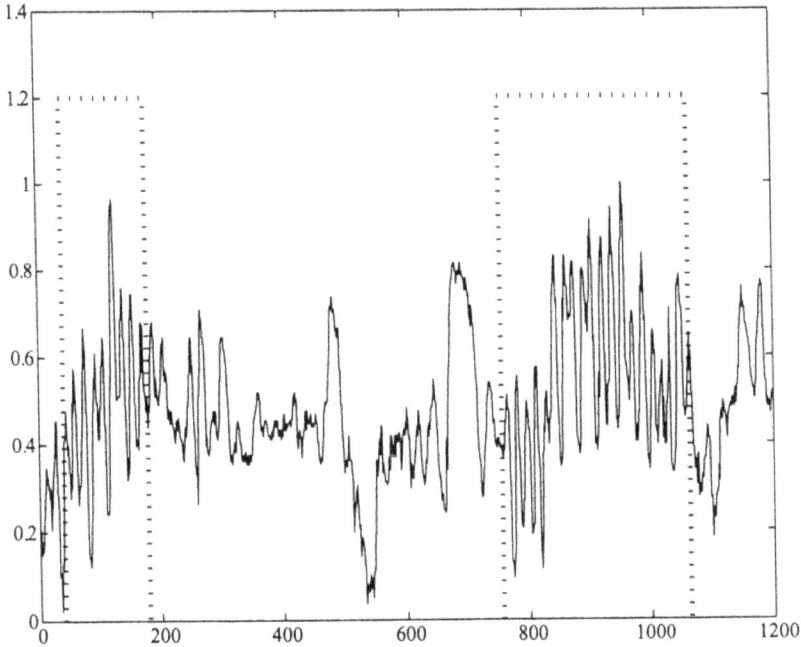

Figure 4.2   The detection result for the segment in Figure 4.1. Two sleep-spindles are detected and marked with dotted lines. The first starts at about the 20th datum (namely: the 0.1th second) with a duration of about 0.8s and the second starts at about the 750th datum (namely: the 3.75th second) with a duration of about 1.5s.

sleep-spindle starting at t_b and ending at t_e. For an automated detection algorithm, the mis-detection degree, simply denoted by MD, is defined as follows: (1) if one sleep spindle is detected from $X(t)$, with starting point t'_b and end point t'_e, then

$$MD = \frac{L_\vee - L_\wedge}{L},\tag{4.1}$$

where, $L_\vee = \max(t_e, t'_e) - \min(t_b, t'_b)$, $L_\wedge = \min(t_e, t'_e) - \max(t_b, t'_b)$ and $L = t_e - t_b$. (2) if no spindle or more than one spindle is/are detected, then $MD = \infty$.

It is easy to see that MD is a nonnegative number and $MD = 0$ if and only if the sleep-spindle is detected accurately, and the smaller MD is, the more accurately the detection does. Table 1 lists the distribution of the MDs produced by Algorithm 4.1 for all the 183 samples and the corresponding histogram is displayed in Figure 4.3, in which all the MDs greater than 1 is included into that of $MD = 1.1$.

It is encouraging to see that most of the MDs are between 0 and 0.2, which shows that our algorithm arrives at satisfying detection results in both locations and durations.

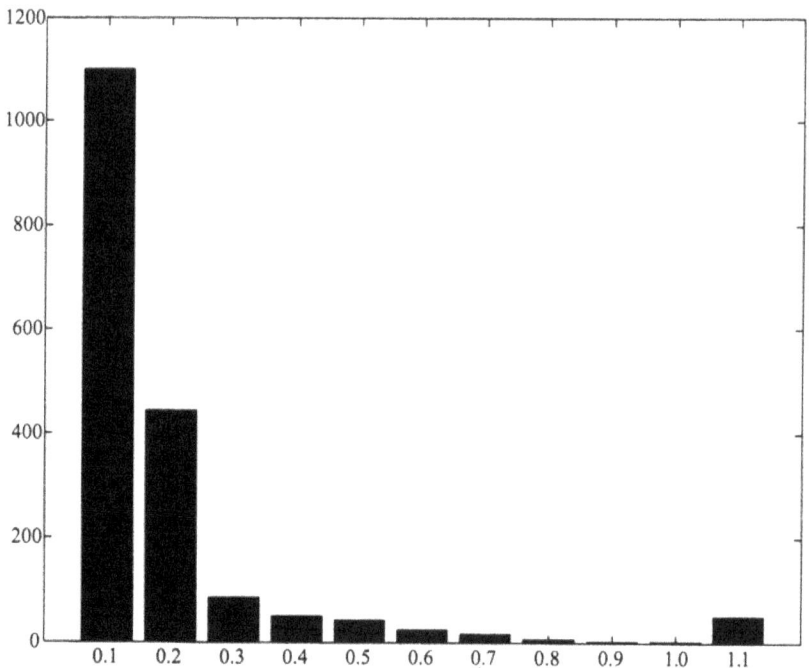

Figure 4.3   The histogram of the MDs corresponding to Table 1.

Table 1   The distribution of the MDs produced by Algorithm 4.1, NS is the number of spindles whose MDs are within the given interval.

MD	0 ~0.1	0.1 ~0.2	0.2 ~0.3	0.3 ~0.4	0.4 ~0.5	0.5 ~0.6	0.6 ~0.7	0.7 ~0.8	0.8 ~0.9	0.9 ~1	>1
NS	1101	443	86	51	43	25	17	7	3	3	51

4.2   Pitch period detection algorithm based on Hilbert-Huang transform

The detection of pitch period from speech signals has been studied by many researches in the past decades. As an important parameter in the analysis and synthesis of speech signals, pitch period information has been used in various applications such as 1) speaker identification and verification, 2) pitch synchronous speech analysis and synthesis, 3) linguistic and phonetic knowledge acquisition and 4) voice disease diagnostics. However, reliable and accurate determination of the pitch period is difficult due to the complexity of the speech signal, which can be viewed as the output of a time-varying system excited by a quasi-periodic train of pulse for voiced speech, or by wideband random noise for unvoiced speech. Therefore, it is still a challenging task to develop algorithms for different applications. Roughly, the techniques that have been developed for automatic detection of the pitch period over the past several years can be classified two categories: (a) event detection pitch detectors, which estimate the pitch period by locating the instant at which the glottis close, and then measuring the time interval between two such events; (b) nonevent detection pitch detectors, which are mainly based on the short-term autocorrelation function and the average magnitude difference function. Generally, the nonevent based pitch detectors are computationally simple. However, they assume that the pitch period is stationary within each segment, so the draw-backs of these techniques are their insensitivity to non-stationary variations in the pitch period over the segment length and unsuitability for both low pitched and high pitched speakers. Comparing with nonevent detection pitch detectors, the event detection pitch detectors are immature. Only a few event based pitch detectors have been developed [2, 3, 9, 16, 41]. Despite the high accuracy, most of them are either applicable to only a part of vowels or of computationally complexity.

Based on Hilbert-Huang transform, a new event detection pitch detector is presented by Yang, Huang and Yang [51]. Because of the high time-frequency local character and being applicable to nonlinear and non-stationary process, HHT is employed to locate the instant at which the glottal pulse takes place. Then, the pitch period is detected accurately by measuring the time interval between two glottal pulses. The algorithm is described as follows [51].

Algorithm 4.2. *Let $x(t)$ be a segment of voiced speech signal.*

Step 1 *Decompose $x(t)$ into IMFs by EMD algorithm;*

Step 2 *For each IMF, Compute its Hilbert transform, instantaneous amplitude and instantaneous frequency respectively;*

Step 3 *Compute $H(\omega, t)$ based on the results of Step 2;*

Step 4 *Compute instantaneous energy*

$$IE(t) = \int H(\omega, t)d\omega;$$

Step 5 *Compute the derivative of IE(t) and denote it as DIE(t);*

Step 6 *Let TH be a given threshold, we process the DIE(t) as below:*

$$\tilde{DIE}(t) = \begin{cases} DIE(t), & \text{if } DIE(t) > TH, \\ 0, & \text{if } DIE(t) \leq TH; \end{cases} \qquad (4.2)$$

Step 7 *Search the local maxima of $\tilde{DIE}(t)$ and the instants at which the local maxima of $\tilde{DIE}(t)$ take place. They correspond to those instants at which the glottal pulse takes place. Finally, the pitch period is detected by measuring the time interval between two glottal pulses.*

To test our algorithm for real speech signals, a segment of a practical speech signal which are recorded in common microphone with sampling frequency of 8000Hz under natural environment is experimented in [51]. It is the speech signal of vowel 'e' spoken by a male speaker. The speech signal, its $D\tilde{I}E(t)$, the detected instants at which the glottal pluses take place and the track of the pitch period are plotted in Figure 4.4 from

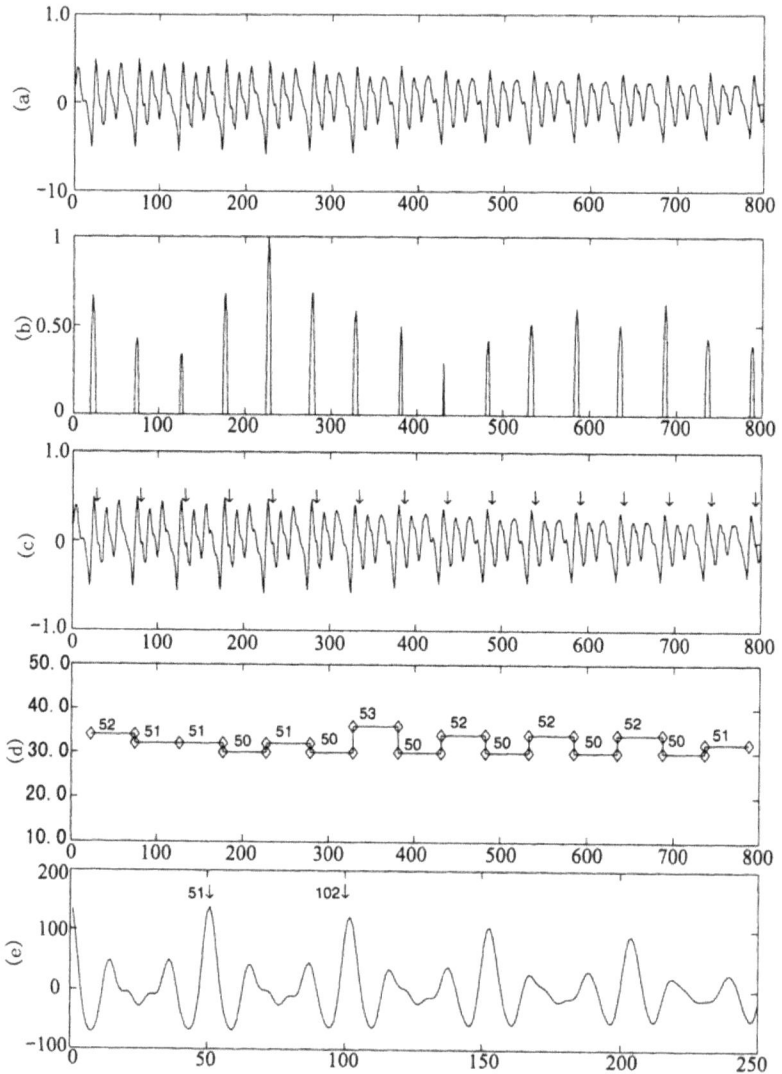

Figure 4.4   The true speech signal of vowel 'e' spoken by a male speaker and the detected result.

top to bottom as marked by (a), (b), (c) and (d). To compare our algorithm with classical algorithms, we also compute the autocorrelation function of the same speech signal and the detected result is plotted on the bottom of Figure 4.4 marked by (e). It is clear that our algorithm exhibits superior performance and high accuracy.

4.3 Chinese font recognition based on empirical mode decompositon

Chinese characters have more than five thousands' history. There are many different fonts and the most frequently used ones are 'SongTi', 'KaiTi', 'FangSong', 'HeiTi', 'LiShu' and 'YouYuan'. Each font has four different styles: Regular, Italic, Bold and Bold Italic. Figure 4.5 shows different font and different style of the Chinese character 'technique', in which from left to right are 'SongTi', 'FangSong', 'KaiTi', 'HeiTi', 'LiShu' and 'YouYuan' respectively and from top to bottom are regular, bold, italic and bold italic styles respectively.

Figure 4.5   Six fonts and four styles of the Chinese character 'technique'.

Font and style recognition is an important and challenging issue in automatic document analysis and processing. It is a necessary process for producing the re-editable text. Meanwhile, if the fonts of a text can be recognized correctly, it can be used to improve the recognition rate of an

OCR (Optical Character Recognition) system. However, partially due to the difficulty of discriminating similar fonts, this topic has been ignored by researchers. Existing studies mainly consist of three classes: (a) the methods based on local attributes such as serifness, boldness, etc. [17, 11]; (b) the methods based on local and/or global typographical features [37, 1, 53, 24]; and (c) the methods based on texture analysis [52, 21]. Different methods employ different kinds of features to recognize fonts. For Chinese texts, because of the structural complexity of characters, font recognition is more difficult than those of western languages such as English, French, Russian etc. Only a few researchers have addressed this issue [52, 20].

Based on Empirical Mode Decomposition, a new method to recognize Chinese fonts was proposed by Yang, Yang, Qi and Suen in [47]. There five basic strokes have been selected to characterize the stroke attributes of Chinese fonts. Based on them, *stroke feature sequences* of a given text block are calculated. By decomposing them with EMD, some Intrinsic Mode Functions are produced and then the first two, which are of the highest frequencies, are used to produce the so called *stroke high frequency energies*, which is the average energy of the two Intrinsic Mode Functions over the length of the sequence. By calculating the *stroke high frequency energies* for all the five basic strokes and combining them with the averages of the five residues, which are called as *stroke low frequency energies*, a ten-dimensional feature vector is formed and used to recognize the fonts and styles.

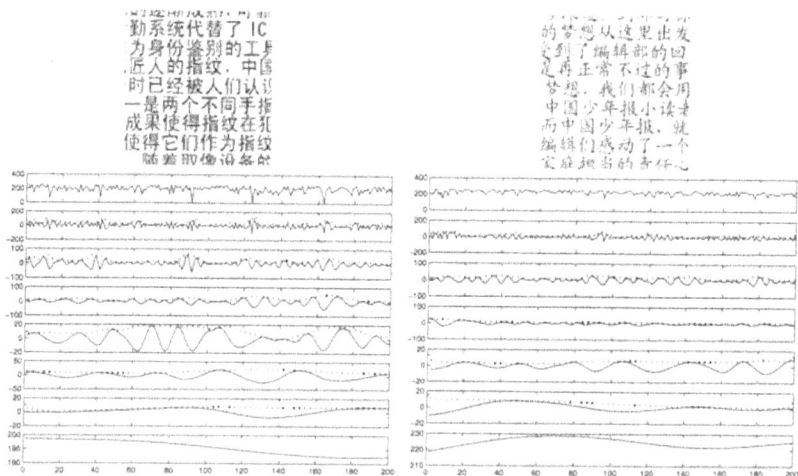

Figure 4.6   Two Chinese text blocks and their EMD decomposition: The left is a text of HeiTi and its 6 IMFs, the right is a text of KaiTi and its 5 IMFs, shown with solid lines. The dotted lines correspond to their instantaneous energies.

In Figure 4.6, the top left shows a Chinese text block of HeiTi of size 128×128; the bottom left shows its *stroke(a) feature series* and its EMD decomposition, in which its 6 IMFs and the residue are plotted with solid lines and their instantaneous energies are plotted with dot lines from top to bottom. The right is the counterpart of the left for a Chinese text block of KaiTi. To observe more clearly, only 200 samples, instead of $N \approx 10 \max(W, H)$, of *stroke(a) feature series* and its IMFs are plotted in Figure 4.6. For simplicity, denote the *stroke(x) feature series* of a text block of HeiTi and KaiTi by S_x^H and S_x^K, their i-th IMF by $\mathrm{imf}_x^H(i)$ and $\mathrm{imf}_x^K(i)$ respectively. Similarly the residues are denoted by R_x^H and R_x^K respectively.

To form the feature vectors, 25 text blocks with the size of 128×128 are selected randomly for each font of FangSong, SongTi, Lishu, KaiTi, HeiTi and Youyuan. Namely, $25 \times 6 = 150$ text blocks are selected. Let N be 1000, the *stroke(a) feature series* is extracted from each text block and decomposed by EMD to produce IMFs. Then the first two IMFs, denoted by $\mathrm{imf}_a^1 = \{\mathrm{imf}_a^1(j)\}_{j=1}^N$ and $\mathrm{imf}_a^2 = \{\mathrm{imf}_a^2(j)\}_{j=1}^N$, are reserved and their total average energy over N is calculated by the following formula:

$$ e_a = \frac{1}{2N} \sum_{j=1}^N [A_a^1(j) + A_a^2(j)], $$

with

$$ A_a^i(j) = \sqrt{[\mathrm{imf}_a^i(j)]^2 + [H(\mathrm{imf}_a^i(j))]^2}, \quad i = 1, 2, $$

where, $H(\mathrm{imf}_a^i(j))$ is the Hilbert transform of $\mathrm{imf}_a^i(j)$. e_a, since it is the high frequency energy corresponding to basic stroke(a), is simply called *stroke(a) high frequency energy*. The EMD residue of stroke(a) feature series is denoted by $\mathrm{res}_a = \{\mathrm{res}_a(j)\}_{j=1}^N$ and its average energy over N is calculated as follows:

$$ r_a = \frac{1}{N} \sum_{j=1}^N \mathrm{res}_a(j), $$

which is the low frequency energy corresponding to basic stroke(a) and, similarly, is called *stroke(a) low frequency energy*.

Conducting the above calculation for all the 125 text blocks we get 125 stroke(a) high frequency energies, e_a, and 125 stroke(a) low frequency energies, r_a, for the six fonts. Each font corresponds to 25 stroke(a) high frequency energies and 25 stroke(a) low frequency energies.

Similarly, 25 text blocks with the size of 128×128 are selected randomly for each styles of SongTi: Regular SongTi, Italic SongTi, Bold SongTi and Bold Italic SongTi. Namely, $25 \times 4 = 100$ text blocks are selected. Let N be 1000, the *stroke(b) feature series*, is extracted from each text block and EMD is imposed on them to produce IMFs. The stroke(b) high frequency energies, denoted by e_b, and the stroke(b) low frequency energies, denoted by r_b, are calculated similarly as above.

Similar discussion can be made for *stroke(x) high frequency energy* and *stroke(x) residue energy* with $x = c, d, and\ e$. Therefore five *stroke(x) energies* together with five *stroke(x) residue energies* corresponding to $x = a, b, c, d$ and e can be calculated. Finally, a ten-dimensional feature vector is produced.

To test our algorithm, experiments have been conducted on six kinds of frequently used Chinese typefaces (SongTi, KaiTi, HeiTi, FangSong, LiShu and YouYuan), each of which includes four styles (Regular, Italic, Bold and Bold Italic), namely a total of 24 classes of fonts. All the samples are grayscale text images of 256 gray levels with a size of 128×128 pixels generated by computer software and by scanner. The computer-generated text blocks are created by Photoshop 7.0 with a resolution of 72 pixels/inch. The scanner-generated text blocks are obtained by an *HP scanjet 3670*, with a resolution of 100 dpi. For each font, 50 computer-generated text blocks (20 of them for training and 30 for testing) and another 50 scanner-generated text blocks (20 of them for training and 30 for testing too) are created. It means a total of 2400 samples for 24 fonts are employed in our experiments.

For the samples given above, the experimental results are shown in Table 2, which gives an average recognition rate of 97.2%. LiShu produces the highest average recognition rate among all six fonts, which is up to 99.1%, and FangSong has the lowest recognition rate, 94.5%. As for the four styles, Regular style has the highest and Bold style the lowest recognition rates, which are 98.6% and 95.7% respectively. The main factor causing the low recognition rate of the Bold style is that the Regular HeiTi is often confused with the Bold SongTi or the Bold YouYuan. As a result, the Regular HeiTi to have the lowest recognition rate of 93.2% among all the Regular fonts.

Table 2   Recognition rate (percent) of fonts and styles.

	SongTi	KaiTi	HeiTi	FangSong	LiShu	YouYuan	Average
Regular	98.2	100	93.2	100	100	100	98.6
Bold	93.2	91.7	100	92.9	100	96.5	95.7
Italic	100	100	96.5	93.3	100	100	98.3
Bold Italic	100	93.3	100	91.7	96.5	94.9	96.1
Average	97.9	96.3	97.4	94.5	99.1	97.9	97.2

Acknowledgement

The author would like to acknowledge all the members of my research group for their joyful collaborations. They are Professors Daren Huang, Zhihua Yang, Ning Bi and my students: Zhijing Yang, Chaoying Zhou, Lihui Tan, etc. Figures 2.2, and 2.3 are from [14], the author thanks The Royal Society press and Professor Norden E. Huang for the permission for use of the these figures.

References

[1] P. Suda, A. Schreyer and G. Maderlechner. Font style detection in documents using textons. *Proc. Third Document Analysis Systems Workshop, Assoc. for Pattern Recognition Int'l*, 1998.

[2] T. V. Ananthapadmanabha and B. Yegnanarayana. Epoch extraction of voiced speech. *IEEE Transacions on Acoust, Speech, Signal Processing*, 23(6): 562–570, 1975.

[3] T. V. Ananthapadmanabha and B. Yegnanarayana. Epoch extraction from linear prediction residual for identification of closed glottis interval. *IEEE Transacions on Acoust, Speech, Signal Processing*, ASSP-27(4): 309–319, 1979.

[4] E. Bedrosian. A product theorem for Hilbert transform. *Proceedings of the IEEE*, 51: 868–869, 1963.

[5] E. J. Beltrami and M. R. Wohlers. *Distributions and the Boundary Values of Analytic Functions*. Academic Press, New York and London, 1966.

[6] H. J. Bremermann. Some remarks on anaytic representations and products of distributions. *SIAM J. Appl. Math.*, 15(4): 920–943, 1967.

[7] J. L. Brown. Analytic signals and product theorems for Hilbert transforms. *IEEE Transactions on Circuits System*, CAS-21: 790–792, 1974.

[8] J. L. Brown. A Hilbert transform product theorem. *Proceedings of the IEEE*, 74: 520–521, 1986.

[9] Y. M. Cheng and D. O'Shaughnessy. Automatic and reliable estimation of glottal closure instant and period. *IEEE Transactions on Acoust, Speech, Signal Processing*, 37(12): 1805–1815, 1989.

[10] L. Cohen. *Time-frequency analysis*. Englewood Cliffs, New Jersey: Prentice-Hall, 1995.

[11] R. Cooperman. Producing good font attribute determination using error-prone information. *Int'l Society for Optical Eng. J.*, 3027: 50–57, 1997.

[12] J. Doman, C. Detka and T. Hoffman et al. Automating the sleep laboratory: Implementation and validation of digital recording and analysis. *International Journal of Biomedical Computing*, 38: 277–290, 1995.

[13] Fei Huang and Chongxun Zheng. Automated recognition of spindles in sleep electroencephalogram base on time-frequency analysis. *Journal of Xi'An Jiao Tong University*, 36(2): 218–220, 2002.

[14] Norden E. Huang, Zheng Shen, Steven R. Long and et al. The empirical mode decomposition and the Hilbert spectrum for nonlinear and non-stationary time series analysis. *Proc. R. Soc. Lond. A*, 454: 903–995, 1998.

[15] Norden E. Huang, Zhaohua Wu and S. R. Long. On instantaneous frequency. In *Workshop on the recent developments of the Hilbert-Huang Transform methodology and its applications*, Taipei, China, March 15th –17th 2006.

[16] Shubha Kadambe and G. Faye Boudreaux-Bartels. Application of the wavelet transform for pitch detection of speech signals. *IEEE Transactions on Information Theory*, 38(2): 917–924, 1992.

[17] S. Khoubyari and J. J. Hull. Font and function word identification in document recognition. *Computer Vision and Image Understanding*, 63(1): 66–74, 1996.

[18] M. J. Korenberg. A robust orthogonal algorithm for system identification and time-series analysis. *Biological Cybernetics*, 60: 267–276, 1989.

[19] Leukel. *Essential of Physiological Psychology*. The CV Company, USA, 1978.

[20] Chen Li and Xiaoqing Ding. Font recognition of single chinese character ba sed on wavelet feature. *Acta Electronica Sinica*, 32(2): 177–180, 2004.

[21] Yuanyan Tang, Li Zeng and Tinghuai Chen. Multi-scale wavelet texture-based script identificat ion method. *Chinese Journal of Computers*, 23(7): 12–18, 2000.

[22] Jianping Liu, Shiyong Yang and Chongxun Zheng. High resolution time-frequency analysis method for extracting the sleep spindles. *Journal of Biomedical Engineering*, 17(1): 50–55, 2000.

[23] J. H. McClellan, R. W. Schafer, and M. A. Yoder. *Signal Processing First*. Prentice Hall/Pearson, 2002.

[24] YongChul Shin, MinChul Jung and S. N. Srihari. Multifont classification using typographical attributes. In *ICDAR99*, pp. 353–356, Bangalore, India, 1999. IEEE Computer Socety Press.

[25] A. Nuttall and E. Bedrosian. On the quadrature approximation for the Hilbert transform of modulated signals. *Proceedings of the IEEE*, 54: 1458–1459, 1966.

[26] M. Orton. Hilbert transforms, Plemelj relations, and Fourier transforms of distribution. *SIAM J. Math Anal*, 4(4): 656–670, 1973.

[27] J. N. Pandey. The Hilbert transform of Schwartz distributions. *Proceedings of the American Mathematical Society*, 89(1): 86–90, 1983.

[28] J. N. Pandey and M. A. Chaudhry. The Hilbert transform of generalized functions and applications. *Can. J. Math.*, 35(3): 478–495, 1983.

[29] J. N. Pandey and M. A. Chaudhry. The Hilbert transform of Schwartz distribtuions II. *Math. Proc. Camb. Phil. Soc.*, 102: 553–559, 1987.

[30] N. Pradhan and P. K. Sadasivan. The nature of dominant lyapunov exponent and attractor dimension curve of eeg in sleep. *Computers in Biology and Medicine*, 26(5): 419–428, 1996.

[31] J. Pricipe, S. K. Gala and T. G. Chang. Sleep staging automation base on the theory of evidence. *IEEE Transactions on Biomedical Engineering*, 36(5): 503–509, 1987.

[32] Tao Qian, Qiuhui Chen and Luoqing Li. Analytic unit quadrature signals with nonlinear phase. *Phisica D*, 203: 80–87, 2005.

[33] W. Rudin. *Real and Complex Analysis*. New Delhi: Tata McGraw-Hill, 2nd edition, 1987.

[34] Shannahoff-Khalsa David S., Gillin J. Christian and et al. Ultradian rhythms of alternating cerebral hemispheric eeg dominance are coupled to rapid eye movement and non-rapid eye movement stage 4 sleep in humans. *Sleep Medicine*, 2(4): 333–346, 2001.

[35] N. Schaltenbrand, R. Lengelle and J. P. Macher. Neural network model: Application to automatic analysis of human sleep. *Computer and Biomedical Research*, 26: 157–171, 1993.

[36] R. C. Sharpley and V. Vatchev. Analysis of the intrinsic mode functions. *Constructive Approximation*, 24: 17–47, 2006.

[37] H. Shi and T. Pavlidis. Font recognition and contextual processing for more accurate text recognition. In *ICDAR'97*, pp. 39–44, ULm, Germany, 1997. IEEE Computer Society Press.

[38] J. Smith, I. Karacan and M. Yang. Automated analysis of the human sleep EEG. *Waking and Sleep*, (2): 75–83, 1978.

[39] E. Stanus, B. Lacroix and M. Kerkhofs et al. Automated sleep coring: A comparative reliability study of two algorithms. *Electroencephalography and Clinical Neurophysiology*, 66: 448–454, 1987.

[40] Elias. M. Stein. *Harmonic Analysis: Real-Variable Methods, Orthogonality, and Oscillatory Integrals*. Princeton University Press, Princeton, New Jersey, 1993.

[41] H. W. Strube. Determination of the instant of glottal closure from the speech wave. *Journal of the Acoustical Society of America*, 56(5): 1625–1629, 1974.

[42] Lihui Tan, Lihua Yang, and Daren Huang. Necessary and sufficient conditions for Bedrosian identity. *Journal of Integral Equations and Applications*, 21 (1), 2009.

[43] Y. Xu and D. Yan. The Bedrosian identity for the Hilbert transform of product functions. *Proceedings of the American Mathematical Society*, 134(9): 2719–2728, 2006.

[44] Lihua Yang and Haizhang Zhang. The bedrosian identity for h^p functions, *Journal of Mathematical Analysis and Applications*, 345: 975–984, 2008.

[45] Lihua Yang and Chaoying Zhou. A distribution space for hilbert transform and applications. *in press*, Science in China Series A: Mathematics, 2008.

[46] Zhihua Yang, Lihua Yang and Dongxu Qi. Detection of spindles in sleep eegs using a novel algorithm based on the Hilbert-Huang transform. In Mang I. Vai Tao Qian and Yuesheng Xu, editors, *Wavelet Analysis and Applications, Applied and Numerical Harmonic Analysis*, pp. 543–559. Birkhauser Verlag Basel/Switzerland, 2006.

[47] Zhihua Yang, Lihua Yang, Dongxu Qi and Ching Y. Suen. An EMD-based recognition method for chinese fonts and styles. *Pattern Recognition Letters*, 27: 1692–1701, 2006.

[48] Zhijing Yang, Chunmei Qing Lihua Yang and Daren Huang. A method to eliminate riding waves appearing in the empirical am/fm demodulation. *Digital Signal Processing*, 2008.

[49] Bo Yu and Haizhang Zhang. The Bedrosian identity and homogeneous semi-convolution equations. *Journal of Integral Equations and Applications*, 20: 527–568, 2008.

[50] Xianda Zhang and Zheng Bao. *Analysis and Processing of Nonlinear Stationary Signals*. Beijing: National defence and industrial publisher, 1998.

[51] Daren Huang, Zhihua Yang and Lihua Yang. A novel pitch period detection algorithm based on hilbert-huang transform. *Lecture Notes In Computer Science*, 3338: 586–593, 2004.

[52] Tieniu Tan, Zhu Yong and Yunhong Wang. Font recognition based on global texture analysis. *IEEE Transactions on Pattern Analysis and Machine Intelligence*, 23(10): 1192–1200, 2001.

[53] A. Zramdini and R. Ingold. Optical font recognition using typographical features. *IEEE Transactions on Pattern Analysis and Machine Intelligence*, 220(8): 877–882, 1998.

www.ingramcontent.com/pod-product-compliance
Lightning Source LLC
Chambersburg PA
CBHW050641190326
41458CB00008B/2372